EXAMPLES
IN
ELEMENTARY ENGINEERING

EXAMPLES IN
ELEMENTARY ENGINEERING

by

DONALD PORTWAY

*Senior Tutor of St Catharine's College and
Lecturer in Engineering at the University
of Cambridge*

CAMBRIDGE

AT THE UNIVERSITY PRESS

1937

CAMBRIDGE
UNIVERSITY PRESS

University Printing House, Cambridge CB2 8BS, United Kingdom

Cambridge University Press is part of the University of Cambridge.

It furthers the University's mission by disseminating knowledge in the pursuit of education, learning and research at the highest international levels of excellence.

www.cambridge.org
Information on this title: www.cambridge.org/9781316611852

© Cambridge University Press 1937

First published 1937
First paperback edition 2016

A catalogue record for this publication is available from the British Library

ISBN 978-1-316-61185-2 Paperback

CONTENTS

Preface   *page* vii

  I   Statics   1

  II Dynamics   12

  III Hydrostatics   29

  IV Structural Problems   34

  V Heat and Heat Engines   43

  VI Magnetism, Statical Electricity and Electro-
      lysis   51

  VII Ohm's Law. Power and Energy   56

  VIII Electromagnetism, Induction and Electro-
      dynamics   66

  IX Direct Current Machines and Simple
      Alternating Current Problems   70

  X Geometrical and Mechanical Drawing   77

  XI Answers and Solutions   85

PREFACE

This book is intended to supply under one cover a fairly comprehensive set of examples for the use of Engineering Students in the early part of their training. Most of the questions have been taken from examination papers which the writer has set for examinations under the auspices of the Admiralty, the War Office and the Oxford and Cambridge Schools Examination Board, the latter mainly in connection with special papers set for Oundle School. These authorities have kindly given leave to use such questions. Much of the work has inevitably been derived from the numerous examples that are current in the Cambridge University Engineering School.

The examples are intentionally very assorted with no kind of grouping, as it is wished to avoid any kind of hint as to the proper avenue of approach. A few of the less elementary questions have however been marked by a star.

No questions on Applied Mathematics have been included as this subject is dealt with fully in so many school text-books, but in view of the essential importance of Machine Drawing to all Engineering Students a few examples with their solutions have been appended, one or two very simple indeed and others which are slightly more advanced.

The answers in many cases have been worked out only to a degree of accuracy obtainable with a 10 inch slide-rule. It has been assumed that students are conversant with ordinary Engineering constants.

The leave of the Controller of His Majesty's Stationery Office has been obtained for the use of certain of the examples and diagrams.

D. P.

1937

I

STATICS

1. Four forces of 3, 4, 5 and 2 lb. respectively act along the sides of a square, following each other in the order given. The sides of the square are 4 ft. Find the magnitude of the resultant force and determine its perpendicular distance from the centre of the square.

2. A body is being dragged along level ground with uniform velocity by a rope inclined at an angle θ to the horizontal. The coefficient of friction between the ground and the body being μ, shew by means of a triangle of forces or otherwise that the tension in the rope will be least when θ is such that $\tan\theta = \mu$. If the body weighs 5 cwt. and $\mu = 0.3$, find the minimum tension in the rope.

3. A regular hexagon $ABCDEF$ is composed of six heavy rods, each of weight W, freely jointed together and suspended from the point A, two light stiff struts BF and CE being inserted to prevent change of shape. Shew that the force in CE is equal to $0.866W$ and that the force in BF is five times as great.

4. In a geared crab the effort required to move a load of 4 tons is 45 lb., while 35 lb. effort is required to move 2 tons. The handle is at a radius of 18 in. and the drum for the load rope is 12 in. diameter. The driver pinions, starting from the one on the handle, have 15, 20 and 25 teeth respectively, and the followers in the same order have 80, 100 and 120 teeth respectively. Make a diagrammatic sketch of the arrangement and plot the load-efficiency graph. Would you expect the machine to overhaul and if so at what load?

5. A body is pulled up a plane, whose angle of inclination to the horizontal is α, by a force parallel to the slope. If the

2 EXAMPLES IN ELEMENTARY ENGINEERING

angle of friction between the body and the plane is ϕ, shew that the efficiency of the system is given by the expression

$$\frac{\tan\alpha}{\tan\alpha + \tan\phi}.$$

What horizontal force is required to push a body weighing 100 lb. up the plane, if ϕ is 20° and α is 30°?

6. The figure shews a temporary luggage lift which is working up and down a vertical column. The weight of the

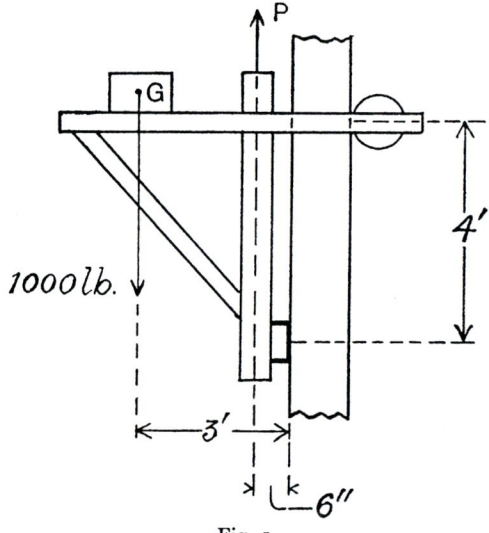

Fig. 1.

carriage and load is 1000 lb., its centre of gravity being at G as shewn. Friction at the wheel is negligible. The coefficient of friction between the slides and the post is 0·25. Find the force P necessary to raise the lift steadily. Make a sketch shewing in magnitude and direction all the forces acting on the lift.

7. Two equal uniform bars of length 16 cm. are freely hinged together. They are placed in a symmetrical position on a rough horizontal cylinder of radius 9 cm. so that each

bar makes an angle of 30° with the horizontal. Find the minimum coefficient of friction between the bars and the cylinder that will allow of equilibrium.

8. A motor-car weighs 2 tons, the distance between its axles being 10 ft. The weight on the back wheels is 25 cwt. when on the level and this increases to 26 cwt. when the front wheels are jacked up 10 in. vertically. Find the height of the centre of gravity of the car above the ground.

9. The figure represents a heavy block of metal *A* supported by a helical spring and guided to move vertically. The weight of *A* is 50 lb., the spring is such that a pull of 50 lb. elongates it 1 in. and the force of friction exerted by the guides resisting up-and-down motion is 10 lb. The weight *A* is pulled down until the total elongation of the spring is 2 in. and it is then released. Prove that it will rise through a distance of 1·6 in. and then drop 0·8 in. before coming permanently to rest.

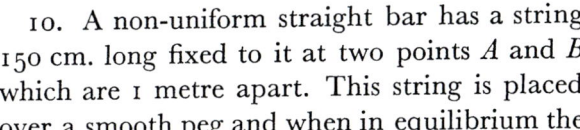

Fig. 2.

10. A non-uniform straight bar has a string 150 cm. long fixed to it at two points *A* and *B* which are 1 metre apart. This string is placed over a smooth peg and when in equilibrium the peg divides the string into two portions 60 cm. and 90 cm. Find the angle that the bar makes with the horizontal and the situation of its centre of gravity between *A* and *B*.

11. A four-wheeled truck has a total distance between its axles of $a + b$. When on a level track its centre of gravity is a distance h above the rails and the centre of gravity is then a horizontal distance a from the front axle and b from the rear axle. If θ is the greatest slope upon which it can stand when both pairs of wheels are locked, find the greatest slope upon which it can stand when only the upper wheels are locked.

12. A camp stool consists of two light bars ABC, DBE pivoted at B, such that $BA = BD = BE = BC$, and at right angles to each other. A weight is placed in the middle of the canvas seat at F, the angle AFD being then 140°. State, giving reasons, whether the feet will tend to approach or recede from each other, and if in this position the state of limiting equilibrium has been reached, find the coefficient of friction between the feet and the ground.

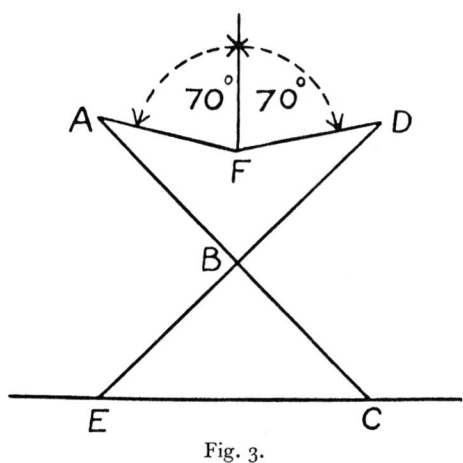

Fig. 3.

13. Two uniform planks of equal length weighing 40 lb. and 50 lb. respectively stand upon a smooth horizontal plane with their upper ends hinged together and are prevented from slipping by a rope joining their feet. If the angle between the planks at the hinge is 60°, what is the tension in the rope and the reaction at the hinge?

14. A simple bridge consists of two similar and uniform girders AC and CB hinged together at C and to their abutments at A and B. The span AB is 12 ft. and the rise of C above AB is 3 ft. Each girder weighs 1 cwt. and a vertical force of 8 cwt. is applied at C. Find

(i) The distance the central hinge will descend if the abutments move 1 in. apart horizontally.

(ii) The magnitude of the horizontal component of the reactions at A and B.

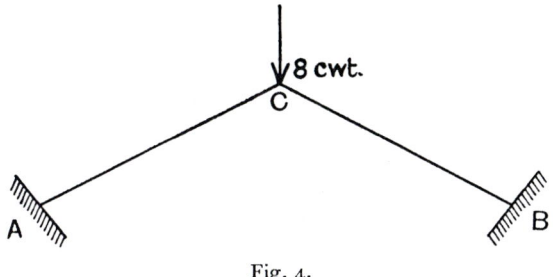

Fig. 4.

15. In the epicyclic train of wheels shewn in the diagram the arm D is capable of rotating about the centre of A. Shew that if A and C have the same number of teeth, C will not rotate on its spindle if A is held fixed and the arm D rotates. If A has 20 teeth and B has 10 teeth, find how many times B rotates for each revolution of D.

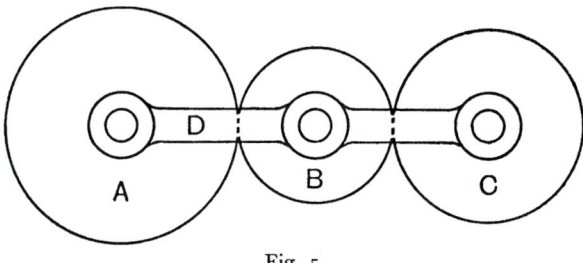

Fig. 5.

16. The countershaft driving a lathe runs at a speed of 180 revolutions per minute. The largest step on the speed cone is 10 in. diameter and the smallest is 4 in. diameter. Each pair of wheels in the back gear has teeth of number 15 and 45 respectively. If the belt is running on the smallest step on the countershaft cone and the back gear is "in", what is the surface speed in feet per min. of a piece of work 7 in. in diameter? What is the greatest possible speed?

17. The figure gives dimensioned side and front elevations of a pair of dockyard sheer-legs, AB and AC, the feet B and C being 20 ft. apart. They are supported by a guy-rope AD and are carrying a load of 15 tons suspended from A. Determine the forces in AB, AC and AD.

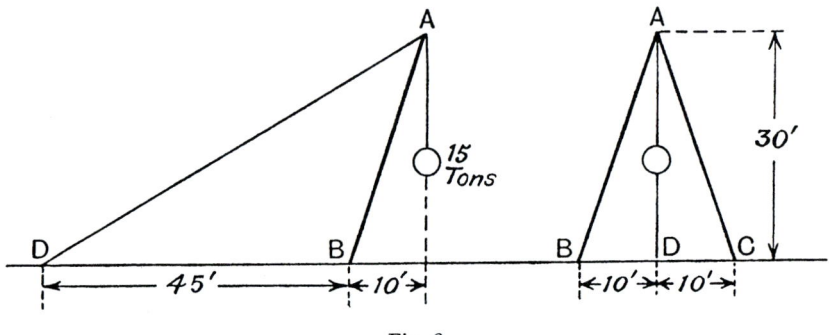

Fig. 6.

18. Six equal rigid bars, whose weight may be neglected, are freely joined at their ends so as to form a regular tetrahedron $ABCD$. This body is suspended from A and three equal weights W are hung from B, C and D respectively. Find the forces thereby produced in all the bars, stating which are in tension and which are in compression.

19. A trunk of pine-wood is 40 ft. long. Its cross-section is circular and it tapers uniformly, its diameter being 3 ft. at one end and 2 ft. at the other. Find the weight of the trunk and the distance of its centre of gravity from the heavier end. Pine weighs 28 lb. per cubic foot.

20. The roadway of a bridge consists of two light girders AB and BC hinged together at B and to the abutments at A and C. The roadway is supported by vertical tie rods FD and GE which are themselves supported by ties HF, FG and GK as shewn. Assuming that the reactions at A, B and C are

vertical, find the tension set up in *FD* and *GE* by a load of 1000 lb. at *P* 12 ft. from *A*.

Find also the tension in *FH*.

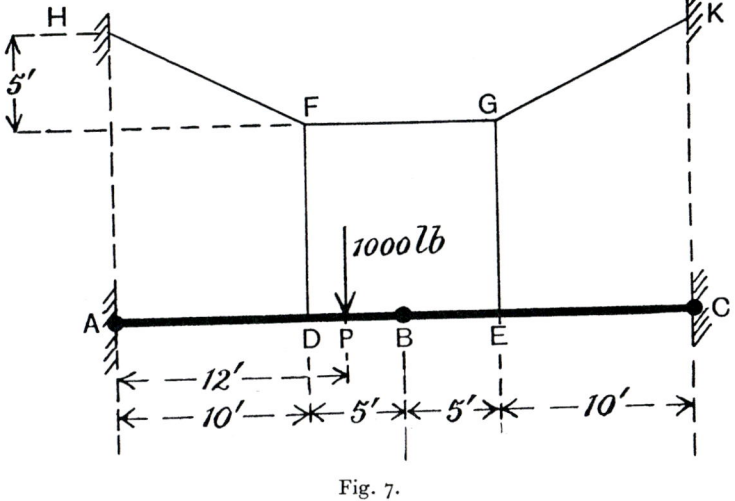

Fig. 7.

21. A bridge girder weighing 3 tons is being hauled across a river in the manner shewn, one winch on the far bank hauling the rope over a derrick tackle with a force *F*, while a winch on the near bank exerts a horizontal force *P* on the

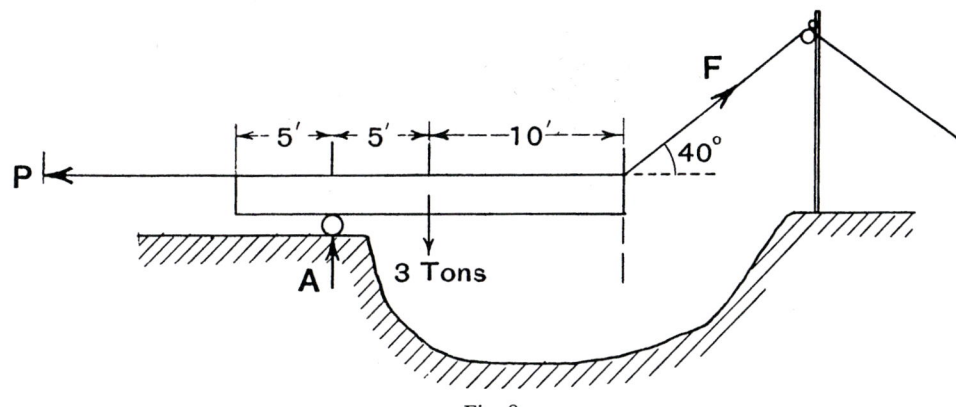

Fig. 8.

preventer tackle. The girder rests on a frictionless roller at A whose reaction is vertical. In the position shewn the girder is in equilibrium and horizontal. Find the values of F and P. Find also the force on the roller.

22. Find the approximate velocity ratio of a Single Spanish Burton consisting of two single blocks and a hook, the rope being rove as shewn. What would happen if the rope were continuous and passed through a pulley at A instead of having two ends, both connected to the hook?

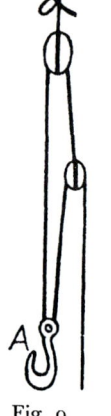

The efficiency of a tackle when lifting 3 tons is 70 per cent. and the velocity ratio is 80. Calculate

(1) The effort required to move this load.

(2) The part of the effort that is then being used in overcoming friction.

(3) The effort that would have to be exerted to prevent the 3 tons from "taking charge".

23. A retaining wall has the form indicated (Fig. 10), the necessary dimensions being given. The wall weighs 40 tons per foot run. A continuous column of water has got between the wall and its backing, shewn shaded from A to D. Find the magnitude of the resultant thrust on the base DE estimated per foot run of wall, and its direction with the vertical.

Fig. 9.

24. Fig. 11 shews two strings passing over two smooth pulleys carrying weights of 3 lb. and 4 lb. A third weight W is suspended from the point joining the two strings and a horizontal force F is applied at this point. If the two strings each make an angle of 50° with the vertical, calculate the values of F and W and check by graphical means.

25. A heavy cylinder of weight W lb. rests on a rough plane and against a rough wall, the coefficient of friction between the cylinder and both ground and wall being 0·2.

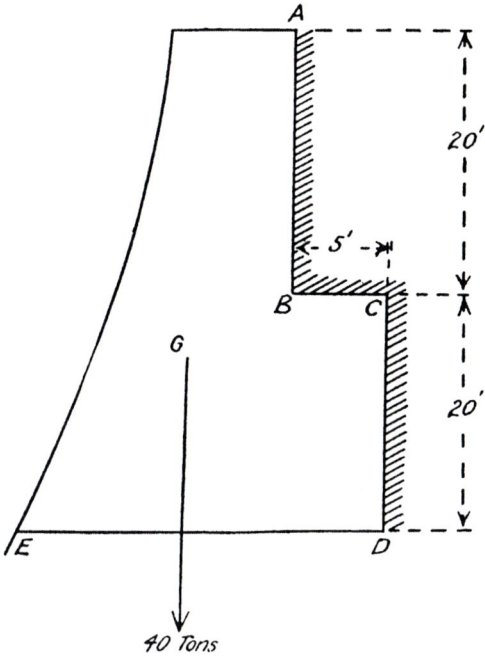

Fig. 10.

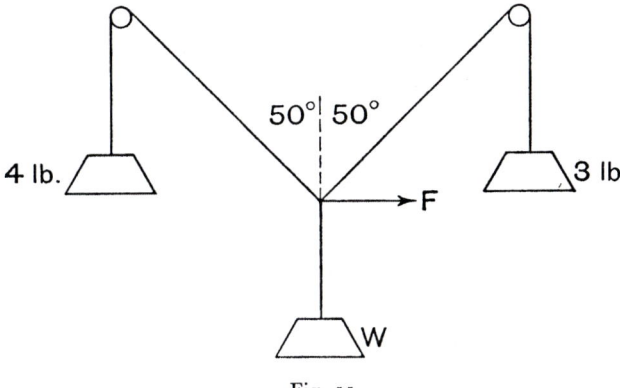

Fig. 11.

It is pushed horizontally at its highest point with a certain force, F lb., until movement just occurs. Shew that this is the case when $F = \frac{3}{11} W$.

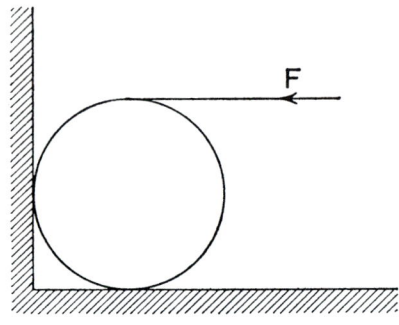

Fig. 12.

26. A tripod is formed of three light bars, each of length l, and from the apex of the tripod a weight W is suspended. The feet of the tripod rest on the ground at the corners of an equilateral triangle of side a. Find the thrust in each rod.

If the feet are just about to slip, what is the coefficient of friction between the feet and the ground?

27. Four equal uniform rods each of weight 5 kilogrammes are hinged together to form a rhombus $ABCD$. The system is suspended from A and is prevented from collapsing by a light rod interposed between B and D and of such a length that the angle at A is 60°. Find the thrust in this rod.

28. A horse has to draw a sledge weighing 2000 lb. up an incline of 1 in 12. The coefficient of friction between snow and runners is 0·05 and the horse can exert a pull of 212 lb. wt. Find the least possible inclination of its path to the line of greatest slope.

29. A wire of uniform section and weight w per unit length is stretched between two points distant l apart at the same

level. The central dip d is very small compared with l. Prove that

(1) The form of the wire is approximately a parabola.

(2) The tension at the lowest point is $w\dfrac{l^2}{8d}$.

(3) The tension at the ends is $\dfrac{wl^2}{8d} + wd$.

If a steel wire may be safely stressed up to 20 tons per sq. in., calculate the longest permissible horizontal span, given that the ratio of dip to span is $\frac{1}{20}$ and that the steel weighs 480 lb. per cub. ft.

II

DYNAMICS

1. A ship is steaming due east at 8 miles per hour and the wind is apparently from S.S.E. The ship then changes course to the S.E. without altering speed and the wind then appears to be due south. What is the speed and what is the true direction of the wind?

2. A steamer is going due north at 12 miles per hour. The wind is due west and the direction of the line of smoke is S. 20° E. What is the velocity of the wind and what is the actual velocity of the smoke as it leaves the ship?

3. An airship is travelling from Cardington near Bedford to Pulham in Norfolk, the distance being 80 miles and the direction 20° N. of E. The wind is from the east and is blowing at 20 miles per hour. If the speed of the airship in still air is 75 miles per hour, what is the least time that the journey can take and in what direction must she steer?

4. A cable drum is of 4 ft. diameter with side flanges of 6 ft. diameter. It rests on its flanges on a rough road and the free end of the cable comes from the underside of the drum and is pulled forward with a constant velocity of 3 ft. per sec. How long will it take to wind up 60 ft. of cable and how far will the drum have rolled in this time?

*5. A road is of width a and a stream of omnibuses of breadth b are following each other at intervals c and moving with velocity V. Shew that the time in which it is possible to cross the road in a straight line with the least uniform velocity is given by the expression $\dfrac{a}{V}\left(\dfrac{c}{b}+\dfrac{b}{c}\right)$, and that this time is never less than $\dfrac{2a}{V}$ whatever the values of b and c.

*6. A solid ball of radius r is projected forward along a rough horizontal plane with velocity v, being given simultaneously an angular velocity ω in a direction contrary to that in which it would roll if there were no sliding. Shew that the ball will cease moving forward and then move backwards if $\frac{2}{5}r\omega - v$ is positive, but that if this quantity is negative it will finally roll in the original direction without sliding.

7. The four-bar chain shewn in the figure consists of two cranks P_1A and P_2B rotating about P_1 and P_2 and pin-jointed at A and at B respectively. The dimensions are as follows:

$$P_1P_2 = 5 \text{ ft.} \quad P_1A = 2 \cdot 4 \text{ ft.} \quad AB = 2 \cdot 5 \text{ ft.} \quad BP_2 = 2 \text{ ft.}$$

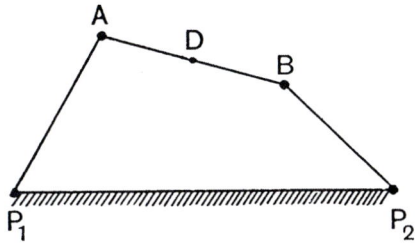

Fig. 13.

If the crank P_1A is moving with constant speed of 60 revolutions per minute, find the velocity of B in ft. per sec. at the instant when the angle AP_1P_2 is 60°.

Find also the velocity of D, the mid-point of AB, in magnitude and direction and the angular velocity of the crank BP_2 at this instant.

*8. Shew that if the retardation of a moving body is proportional to the square of its velocity its motion never absolutely ceases.

A ship has its engines stopped when its speed is 10 ft. per sec. and 1 minute later its speed is 6 ft. per sec. For speeds of 2 ft. per sec. and over it may be taken that its retardation is proportional to the square of its speed. Find how far the ship will drift while its speed falls from 10 ft. per sec. to 2 ft. per sec.

9. If the resistance of the atmosphere and the shape and motion of the earth are ignored, what is the least initial velocity of a shell that is to fall 75 miles from the gun that fires it? Taking the velocity of sound as 1100 ft. per sec., examine whether the shell would arrive before the report from the gun. Indicate briefly the effect of the factors ignored above on the range of the shell.

10. An empty box whose height is h and whose bottom is a square of side a stands on a level road. The coefficient of friction between box and road is μ. The wind is blowing perpendicularly to one vertical side producing a force that may be assumed to act at the centre of the side and which just causes the box to move. Shew that it will slide across the road if μ is $< a/h$, but that if μ exceeds this value it will be blown over.

11. A soldier is using a rifle whose muzzle velocity is 1000 ft. per sec. He fires with accurate aim at a mark which he estimates as 500 yd. away. Actually it is 400 yd. away. By how much will he miss it? Assume that the trajectory is parabolic.

12. A body whose mass is 64 lb., initially at rest, is acted upon by a force whose magnitude is $2t + 5$ lb. weight, where t is the time in seconds that has elapsed from the start. Find the velocity and acceleration of the body after 6 sec. and the distance it has then travelled.

13. A soldier throws a bomb at an angle of 30° with the horizontal and with an initial velocity of 80 ft. per sec. If the bomb explodes 2 sec. after he releases it, find, without use of formula, how high above him it will be when it goes

off and what horizontal distance it has travelled. Air resistance may be neglected.

How long will it be before the bomb has reached its highest point?

14. The mass of a gun is 20 tons and the mass of the shell is 800 lb. The muzzle velocity of the shell is 2800 ft. per sec. and the recoil is resisted by a constant force of 200 tons. Calculate the distance through which the gun will recoil.

15. The velocity of a body increases uniformly with the distance travelled and is 20 ft. per sec. when it has moved 100 ft. What is its acceleration after travelling 100 ft. and how long does it take to traverse from 50 to 100 ft. from the start?

*16. A chain which is 20 ft. long and which weighs 4 lb. per ft. lies coiled on the ground. One end is raised vertically with a constant acceleration of 1 ft. per sec. per sec. Find the lifting force when x ft. of chain have left the ground, and hence determine the total work that has been done by the lifting force by the time that the chain is just clear of the ground.

17. A motor-car engine gives out its maximum horse-power at 2400 revolutions per minute, its horse-power then being 25. The back wheels have a diameter of 30 in. The weight of the car is 1200 lb., the wind resistance being $v^2/80$ lb., where v is the velocity in ft. per sec. If the ratio of engine revolution to road wheel revolution is $\rho : 1$, write down the cubic equation for determining the value of ρ that will enable the car to climb a hill of 1 in 10 ($\sin^{-1} \frac{1}{10}$) when developing its maximum horse-power. Resistances other than wind and gravity are to be neglected.

18. A uniform rod 3 ft. long is set rotating about its midpoint on the surface of a rough horizontal table with angular velocity 120 revolutions per minute. If the coefficient of friction between rod and table is 0·2, find how many seconds

it will be before it comes to rest. Find also how many revolutions the rod will make before its speed of rotation is halved.

19. The coefficient of sliding friction between the wheels of a car and the road is 0·6. If the clutch is withdrawn and the brakes are applied so as to lock the wheels when the car is travelling on the level road at 54 miles per hour, find the distance travelled and the time taken before it comes to rest.

What assumptions are necessary in your calculations?

20. A locomotive of mass 80 tons develops a horse-power of 1000 when drawing a train of mass 200 tons. If the resistance to motion is $V/2$ lb. per ton, where V is the speed in miles per hour, what is the greatest velocity up a long slope whose gradient is 1 in 100?

21. A tank whose weight is 40 tons is climbing a bank which rises 1 ft. for every 4 ft. along the slope. It is assisted by a stationary tank at the top which is pulling on a wire rope parallel to the slope with a constant force of 2 tons' weight. The tank itself is moving up with a steady velocity of 2 ft. per sec. but the driving bands owing to skidding are actually moving downhill at $\frac{1}{2}$ ft. per sec. Calculate (a) the coefficient of sliding friction between bands and ground, (b) the total horse-power transmitted to the bands.

Discuss whether the tank can be expected to climb unaided a hill on similar ground of 1 ft. rise for 5 ft. along the slope.

22. A truck whose weight is 2000 lb. is moving with a velocity of 10 ft. per sec., the propulsive power applied being constant and equal to 2 horse-power. Frictional resistance may be neglected. Find the time that will elapse before the acceleration is reduced to one-half of its initial value. Find also the distance that is covered by the truck in this time.

23. Find what extra horse-power must be exerted by a locomotive to keep up a speed of 75 ft. per sec. during the time that it is picking up 8 tons of water at a uniform rate from a trough situated between the rails and 400 yd. long.

The work required to lift the water from the troughs into the tender may be neglected.

24. A bullet weighing $\frac{1}{2}$ oz. is fired horizontally into a block of wood weighing 50 lb. which is suspended by a cord on a pendulum. The bullet remains embedded in the wood, which swings a distance of 4 in. before coming to rest. The periodic time of swing is found to be 2·5 sec. Determine approximately the velocity of the bullet, stating what assumptions and approximations are made in the calculations.

25. A motor-car has a wheel base of 8 ft. Its centre of gravity is 3 ft. from the ground and three-fifths of the whole weight is carried by the back wheels when the car is at rest on level ground. Find the least distance in which it can be stopped by brakes on the back axle alone from a speed of 30 ft. per sec., the maximum coefficient of friction between the tyres and the ground being 0·8. Neglect any effect due to the rotational inertia of the wheels. The ground reaction at the front wheels may be assumed to be vertical.

26. A body starts with a velocity of 100 ft. per sec. and moves in such a way that its retardation is equal to one-quarter of the distance in feet traversed. Find how far it moves before it stops.

27. The service field gun weighs 24 cwt. and fires an 18 lb. shell with a muzzle velocity of 1600 ft. per sec., but after 2000 yd. this velocity has decreased to 1000 ft. per sec. Calculate (a) the initial velocity of recoil of the gun, (b) the average resistance of the air during the first 2000 yd. of the shell's motion.

28. Two trucks are running on a level track in the same direction; the front one weighs 10 tons and is moving at a speed of 2 ft. per sec., the rear one weighs 6 tons and is moving at 5 ft. per sec. Each truck has two buffers each

requiring 3 tons to compress it 1 in. Find the common velocity at the moment the springs are both compressed to the maximum extent. Find also the amount the springs compress on the assumption that 20 per cent. of the energy otherwise available is dissipated in friction during compression.

29. The following figures are taken on a level road in a motor-car test:

Velocity in miles per hour	10	20	30	40	50	55	60
Time in seconds	0	6	13	20	31	42	59

Find how far the car has travelled in the 59 sec.

Taking the weight of the car as 1 ton and road and wind resistance as 40 lb., find the horse-power required to overcome this resistance and to produce the acceleration when travelling at 40 miles per hour in the above test. The rotational inertia of the wheels may be neglected.

30. A lorry whose loaded weight is 4 tons is running at a given moment at 12 miles per hour up a slope of 1 in 56, road resistance being assumed constant and equal to 25 lb. per ton, the horse-power effectively employed being then 12. Find its acceleration, and assuming that resistance and horse-power both remain unchanged, find the speed up the slope when the acceleration has fallen to half its former value.

Criticise briefly the assumptions in this problem.

31. The table gives the effective tractive force acting on a motor-car weighing 1000 lb. at the given distances from the start. Frictional resistance may be taken as constant and equal to 20 lb. weight throughout. Plot the velocity distance graph and find the acceleration of the car at 60 ft. from the start.

Force in lb. weight	100	90	80	70	60	50
Distance from start in feet	0	20	40	60	80	100

32. Two masses, of 20 lb. each, are joined together by a string 4 ft. long, and are placed in contact with each other on a rough horizontal table, the coefficient of friction between table and masses being 0·2. A string from the mass nearer the edge of the table is placed over a pulley at the edge, and at the other end of this string is a weight of 5 lb. If this latter weight is let go, find (a) the maximum velocity of the hanging weight; (b) the velocity of the system just after the string between the two masses has tautened, (c) the tension in this string during the motion after it has tautened, (d) the total time during which there is motion.

33. A machine gun is mounted on an aeroplane, and when the latter is travelling at 50 miles per hour the gun is fired in the direction of travel for 15 sec. Find the reduction in speed of the aeroplane due to this and the force tending to move the gun relative to the aeroplane. The total weight of the aeroplane is 1800 lb. The gun fires 600 bullets per minute, each weighing $\frac{1}{2}$ oz., with a muzzle velocity of 1000 ft. per sec.

34. A cruiser is propelled at 25 knots by means of engines developing a shaft horse-power of 50,000, and running at 150 revolutions per minute. Calculate the torque given to the propeller shaft, and the resistance to motion.

Assuming that the resistance varies as the square of the speed, what horse-power would be required at 30 knots?

The propulsion efficiency can be taken in both cases as 65 per cent. (1 knot = 6080 ft. per hour.)

35. In a tennis court the service line is 39 ft. from the net, which is 3 ft. high. If a ball is served horizontally 7 ft. from the ground so as just to clear the net, how far from the net on the other side will it strike the ground and with what velocity is it served? (Neglect air resistance.)

36. A 10 ton tank moving at 16 ft. per sec. fires two machine guns at an enemy straight ahead, each gun firing ½ oz. bullets at the rate of 600 per minute. The muzzle velocity of the bullets being 1000 ft. per sec., find the retardation produced on the tank if the engine continues to exert the same propulsive force on the tank. Find also the extra horse-power that the engine must exert to maintain its speed.

37. Two railway trucks weigh 4 tons each and are coupled together. Starting from rest they run down an incline of 1 in 100 and 500 yd. long. The frictional resistance is constant throughout and equals 10 lb. weight per ton. Each truck has four wheels each weighing 400 lb. and they are solid circular discs of 3 ft. diameter. The moment of inertia of the axles is negligible. Find the speed at the bottom of the incline, allowing for the rotational inertia of the wheels.

If the trucks are then brought to rest in 200 yd. by steady braking of the rear truck, find the tension in the coupling during this process.

38. Two masses, M_1 and M_2, are tied together by a string which passes over a light pulley at the apex of a wedge, as shewn in the sketch. If the coefficient of friction between the masses and the wedge is 0·3, find the ratio of M_1 to M_2 if M_1 just descends.

By considering the forces acting on the wedge investigate whether it has any tendency to move, and, if so, in which direction.

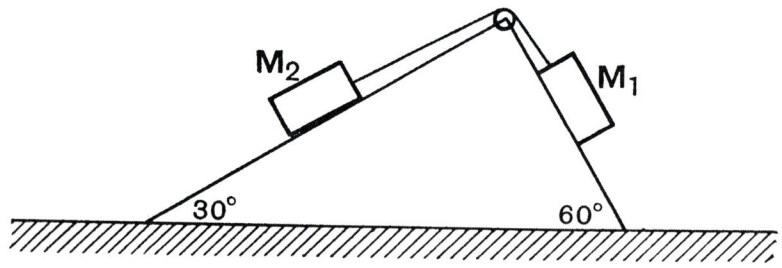

Fig. 14.

39. A chain, 10 ft. long, and weighing 1 lb. per ft., is laid on a rough table, so that part of it hangs over the edge, the coefficient of friction between chain and table being 0·25. Find how much will hang over the edge so that equilibrium is just maintained. Find also the acceleration of the chain at the moment when it has slipped through a further 3 ft.

40. A tap has a nozzle of 1 cm. diameter. It is turned on and the water leaves the mouth of the tap with a vertical velocity of 5 metres per sec. The water falls into a bucket weighing 1 kilogramme placed so that its bottom is a metre below the nozzle. The bucket is standing on a spring balance. What would the balance read just after turning on the tap and before water had time to accumulate in the bucket?

41. A gun fires a shell of mass 850 lb. with a muzzle velocity of 2500 ft. per sec., the relevant radius of gyration of the shell being 4 in. The length of the barrel is 50 ft. and the shell makes two complete revolutions before leaving the muzzle. Calculate the total kinetic energy of the shell as it leaves the gun and find the couple acting about the axis of the barrel during discharge, the acceleration of the shell being assumed uniform.

42. In a cement works the cement, falling in a steady stream, drops with a velocity of 32 ft. per sec. on to a trap-door which is controlled by a spring which is such that it deposits 380 lb. of cement into a sack under the door every time the force on the spring reaches 400 lb. What time elapses between the closing and the re-opening of the trap-door?

43. An aeroplane flies a course in the form of a square. Its air speed is 120 miles per hour and there is a 30 miles per hour wind blowing along a diagonal. Neglecting any loss due to cornering, calculate the average speed for the course.

*44. If the resistance to the motion of a vessel varies as v^n, where v is the velocity through the water, shew that the rate of propulsion against a tide of velocity a, so that the work done by the engines in a given journey may be a minimum, is $\dfrac{n+1}{n}\,a$.

*45. An electric train has an acceleration on the level which is constant up to 30 ft. per sec., which it reaches in $13\frac{1}{3}$ sec. Above this speed its acceleration may be expressed as $\dfrac{(90-v)^2}{1600}$ ft. per sec., where v is the speed in ft. per sec. How long will it take from rest to reach a speed of 75 ft. per sec. and how far will it travel in doing so?

46. A body has a constant resistance to motion of 12 lb. per ton. It is subject to a force which increases uniformly from zero for 30 sec. and then decreases to zero at the same rate. If the greatest value of the force is such that the velocity of the body is the same at the beginning and end of the time, find the greatest change of velocity of the body.

47. A heavy uniform square plate whose sides are 48 cm. in length is pivoted at one corner so that it can move in a vertical plane. The plate is released when the centre of gravity is on the same horizontal level as the pivot. Find its greatest angular velocity in the subsequent motion.

48. Assuming that g varies inversely as the square of the distance from the centre of the earth, find approximately the height of a mountain which is such that a pendulum that beats seconds at sea-level loses 10 sec. a day. Take the radius of the earth as 4000 miles.

49. An elastic string whose unstretched length is 2 ft. has a weight of 10 lb. attached and makes 60 revolutions per minute as a conical pendulum. The string is then 2 ft. 3 in. long. Find
   (i)   The angle made with the vertical by the pendulum.
   (ii) The tension in the string.
   (iii) The kinetic and potential energies of the system.

50. A pendulum is hung from the roof of a railway carriage. At what angle with the vertical will it hang (1) when the train is running on a curve of 800 ft. radius at a constant speed of 40 miles per hour, (2) at the instant when the train is running on the same curve at the same speed but decelerating at 2 ft. per sec. per sec.?

51. A string of length l is attached at one end to a fixed point P, there being a small body of mass m at the other end. The string is extended horizontally and the mass is then liberated. When the string has swung into the vertical position it encounters a peg Q distant a vertically below P and the mass then performs a circular motion about Q. Calculate the tension in the string just before and just after the peg is encountered.

52. A bottle weighing a kilogramme is floated in a liquid whose specific gravity is 0·8. The cross-sectional area of the bottle at and about the level of the liquid is 5 sq. cm. Shew that its oscillations are simple harmonic when it is disturbed and find their periodic time.

53. A U tube of uniform bore contains a total length of 10 in. of liquid. Calculate the periodic time with which the liquid oscillates when disturbed.

54. A pile of mass 700 lb. is being driven into the ground by a "monkey" of mass 100 lb., which falls 6 ft. on to the top of the pile and does not rebound. If the resistance of the ground may be assumed to be constant and equal to 2400 lb. wt., how far will the pile be driven in by one blow?

55. A grandfather clock is intended to be made so that the length of its "simple equivalent pendulum" is 90 cm., but owing to an error it is actually 3 mm. more than this. How many minutes a day will the clock gain or lose?

56. A particle is held inside and on the rim of a smooth hemispherical bowl. It is allowed to roll down to the bottom of the bowl. Shew that when the particle reaches the lowest point of the bowl the pressure between the bowl and the particle is three times the weight of the particle.

57. A motor lorry is travelling along a road which is worn into sinusoidal undulations. The wave-length is 3 ft. and the height of the crests above the mean level is $\frac{1}{2}$ in. For the rear axle the unsprung mass is $\frac{1}{2}$ ton and the spring-borne mass is 2 tons. Assuming that this latter exerts a constant downward force on the axle, find the critical speed at which the axle begins to leave the road.

58. Water is flowing along a horizontal pipe with two right-angle bends in it as shewn. The pipe is uniform and of

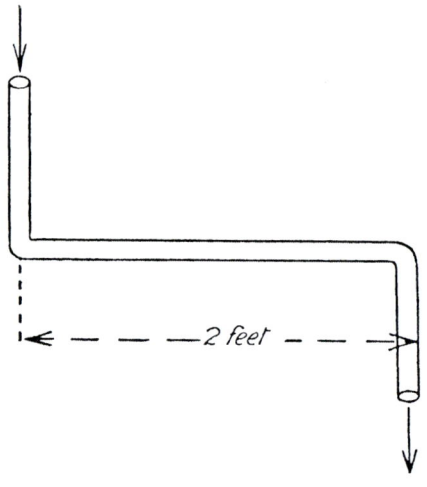

Fig. 15.

6 sq. in. cross-sectional area, and the water is flowing at 20 ft. per sec. Calculate the torque exerted on the pipe.

*59. A body is thrown vertically upwards with velocity V, the resistance being equal to KV^2, where K is a constant.

Shew that the height reached is $\dfrac{1}{2K}.\log_e\left(1+\dfrac{KV^2}{g}\right)$, and express in the form of an integral the time taken to reach this height.

*60. A jet of water moving with a velocity of v ft. per sec. strikes the buckets of a Pelton wheel, the cross-section of one of which is shewn, and which are moving in the same direction as the jet with a velocity of u ft. per sec. Shew that the maximum theoretical efficiency of the wheel is $\dfrac{1+\cos 20^\circ}{2}$, and that this occurs when $u=\dfrac{v}{2}$.

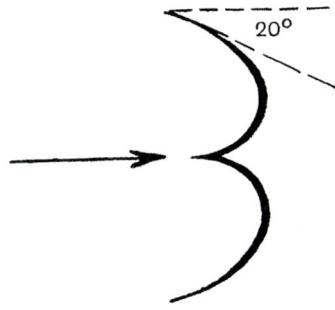

Fig. 16.

What is the theoretical efficiency when the bucket is shaped so that the angle is 0° instead of 20°, and when the wheel is moving with one-third the velocity of the jet?

61. A solid cylinder 20 in. high and 2 in. in diameter stands upright on the floor of a railway carriage, which is travelling round a curve of 800 ft. radius. If the floor remains horizontal, find the greatest speed at which the cylinder can stand upright, and explain why equilibrium can no longer be maintained if this speed is exceeded.

62. An electric train having a mass of 200 tons starts from rest. The initial draw-bar pull is 6·4 tons and the current is regulated so that the pull decreases uniformly with the time until it becomes zero at the end of 2 min. Neglecting wind

and other resistances, calculate the speed at the end of 2 min. and the distance run in this time.

63. An equilateral triangle of side 10 cm. is cut from sheet metal weighing 20 kg. per sq. metre. Find the moment of inertia of the triangle about one side.

64. Find the moment of inertia of the given cross-section of a cast-iron standard about an axis AA' through its centre of gravity.

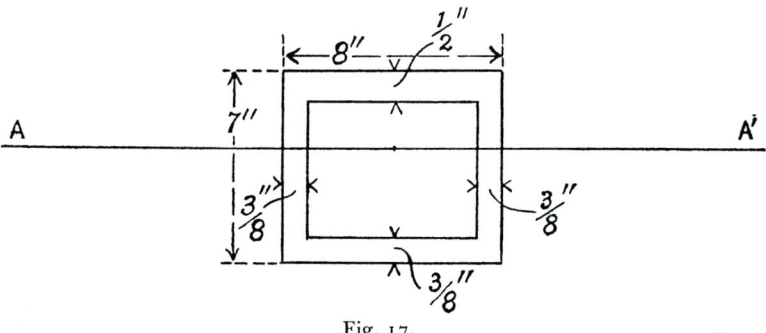

Fig. 17.

65. The diagram gives the relationship in an engine between the turning moment on the crank shaft and the angle turned through by the crank for all positions throughout a complete revolution. The mean speed is 180 revolutions per minute and the engine is driving a dynamo giving a constant resisting moment.

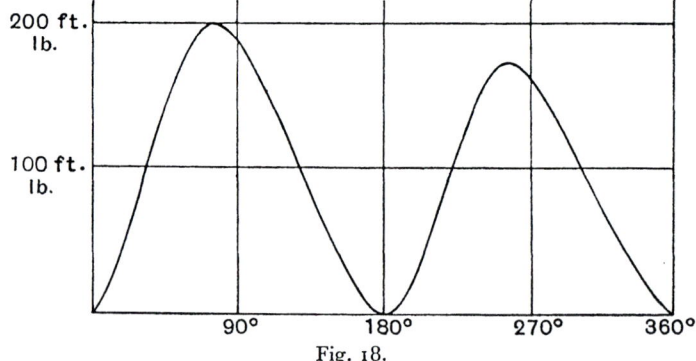

Fig. 18.

Calculate the approximate moment of inertia of the flywheel so that the fluctuation of speed should not exceed 2 per cent. The inertia of the other moving parts may be neglected.

66. A bullet of mass m is fired into a fixed cubical block of mass M so as to strike one of its faces normally in the middle and it penetrates to a depth x. Shew that if the block be free to slide on a smooth horizontal plane the bullet fired as before will penetrate a depth $\dfrac{xM}{M+m}$.

67. A shell has a velocity of 1200 ft. per sec. and bursts into a great number of fragments of equal mass, the velocity impressed on each fragment by the bursting charge being 200 ft. per sec. Find the angle of the cone of dispersion of the fragments.

68. An engine of mass 100 tons and capable of 1200 horse-power draws a train of mass 300 tons. If the resistance to motion is $V/2$ lb. per ton, where V is the speed in miles per hour, find the greatest speed up a bank of 1 in 100.

69. Shew that if a straight tunnel were bored from London to Antwerp, a distance of 180 miles, a train would traverse it under gravity alone in about 42 min. and that its maximum velocity would be about 400 miles per hour. The radius of the earth may be taken as 4000 miles.

How would you expect the time taken to alter for a longer tunnel?

70. A crank AB of radius r revolves about A with constant angular velocity ω, BC being a connecting rod of length l with the end C moving in a straight line through A. If θ is the angle BAC and if the ratio $\dfrac{r}{l}$ is small, prove that the acceleration of C is given very approximately by the expression

$$r\omega^2 \cos\theta + \frac{r^2\omega^2}{l} \cos 2\theta.$$

71. A uniform rod can turn freely in a vertical plane about a point distant one-quarter of its length from its upper end. What must be the length of the rod in order that each swing may take 1 sec.? Why will the periodic time vary if the amplitude of the swing is made considerable?

72. A ship A of 2000 tons is towing another B of 1000 tons. The hawser is such that a tension of 10 tons elongates it 1 in. per 100 ft. At the instant the hawser becomes taut the velocities of A and B are 10 ft. and 7 ft. per sec. respectively. Find the velocity common to the two ships when the hawser is stretched to its maximum, and if l is the length of the hawser in feet, prove that the tension momentarily produced is $1500/\sqrt{l}$.

III

HYDROSTATICS

1. A vessel displacing 5000 tons anchors in a river at high water, when the water may be assumed to be salt (64 lb. per cub. ft.). What weight must be taken out of the ship in order that she may float at the same draught at low water, when the water may be assumed to be fresh?

2. From one end of a balance is hung a solid lump of iron and from the other is hung a solid lump of copper, so that each hang totally immersed in a vessel of oil weighing 40 lb. per cub. ft. and balance each other. The specific gravities of iron and copper are 7·8 and 8·8 respectively and the volume of the lump of iron is 10 cub. in. Find the volume of the lump of copper.

3. A raft is made by fixing a superstructure over four equal hollow cylindrical drums each 10 ft. long and 4 ft. in diameter, their axes being horizontal. It then floats in sea water so that the lowest portions of the drums are 6 in. under water. What weight on the raft will just immerse the drums?

4. A ship displaces 10,000 tons, her total length being 360 ft. and her longitudinal metacentric height 300 ft. Her draught is 21 ft. forward and 22 ft. aft. Find the new draught forward and aft consequent on pumping 60 tons of oil 150 ft. forward. The change of draught at each end may be assumed the same.

5. Shew that if a ship is heeled through an angle of θ radians by a weight of w lb. moved d ft. across the deck and if W lb. is the total displacement of the ship, then the transverse metacentric height is given by the expression $wd/W\theta$ ft. Hence find the angle of heel in degrees caused by a boat weighing 5 tons being moved 20 ft. from one side to the other

of a ship displacing 1000 tons, whose transverse metacentric height is 2 ft.

6. A hydrometer for measuring the specific gravity of the acid in an accumulator consists of a graduated glass tube, of uniform cross-section, loaded with shot below. The tube has a diameter of $\frac{1}{2}$ cm., and rises out of the liquid by 1 cm. as the specific gravity of the acid increases from 1·18 to 1·20. Calculate the weight of this hydrometer in grammes.

7. A uniform spar 10 ft. long floats in the sea and one end is raised $3\frac{1}{2}$ ft. above the surface of the water. If the wood weighs 32 lb. per cub. ft., shew that the other end will sink about $1\frac{1}{2}$ ft. below the water line.

8. A covered-in pontoon with vertical sides and ends and rectangular cross-section is 10 ft. long, 4 ft. wide and 2 ft. deep, and is floating in sea water. A vertical load P is gradually applied to one edge, until one side is completely immersed in the water. The pontoon itself is symmetrically built, so that its weight is central. Find the weight of the pontoon and the magnitude of the applied load P.

9. A cork weighing 1 oz., and whose density is one-quarter that of water, is suddenly freed from the bottom of a tank of water 4 ft. deep. Find with what velocity it will reach the top (1) neglecting the resistance of the water, (2) assuming that the resistance offered by the water is proportional to the distance of the cork from the bottom of the tank, and that it is equal to 2 oz. at the instant that it reaches the top.

From what source is this kinetic energy derived?

10. The vertical cross-section of a concrete dam is a right-angled triangle, the water side being vertical and 13 ft. high. The horizontal base is 10 ft. thick and there is a depth of 12 ft. of water. Consider 1 ft. length of the dam and calculate the

thrust on this length due to the water and the overthrowing moment about the base.

If the concrete weighs 150 lb. per cub. ft., determine at what point in the base the resultant force on the dam acts.

11. A water tank is 8 ft. long, 4 ft. wide and 3 ft. deep. It is completely closed except for a pipe 1 in. in diameter which runs vertically upwards out of the tank and is open to the air. The tank is full of water and the pipe contains water up to a level of 2 ft. above the top of the tank. Find

(a) The total pressure on one end due to the water.

(b) The position of the corresponding centre of pressure.

(c) The total pressure on the bottom due to the water.

(d) The total weight of water.

Account for the difference between (c) and (d). Disregard atmospheric pressure.

*12. The righting couple tending to return a sailing boat to an even keel is proportional to $\sin \theta$, where θ is the angle of heel. The overturning couple due to a wind of given velocity can be assumed to be independent of the angle of heel. A wind which would produce a steady heel of 12° suddenly acts upon the boat. Shew that it will heel to an angle ϕ, which is given by the equation

$$\cos \phi + \phi \sin 12° = 1,$$

and find the approximate value of ϕ.

*13. A rectangular pontoon 35 ft. long and 16 ft. broad carries a crane. They together weigh 64 tons and their centre of gravity is 2·5 ft. above the bottom of the pontoon. Find the metacentric height for rolling when floating in the sea.

If the crane moves a load of 1 ton from amidships transversely through 12 ft., how much will the pontoon heel over?

14. A hydraulic safety valve consists of a conical valve C of weight 20 lb. and exposing a surface of 20 sq. in. to the

water. To what height h above AB must water rise in the
vertical arm in order that the valve may lift?

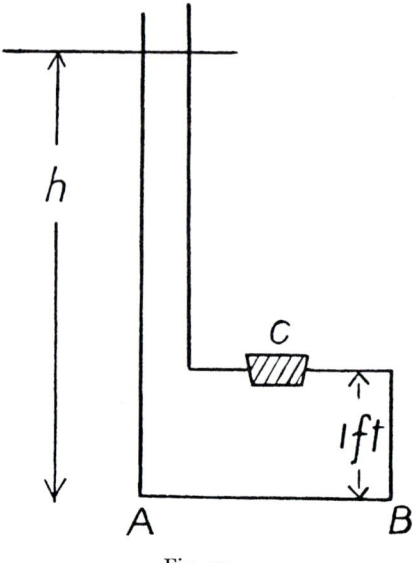

Fig. 19.

15. An assault bridge is made up of empty petrol tins,
there being 18 two-gallon tins to each float with a superstructure above the tins, the weight of each float being 90 lb.
The floats may be submerged till the top of the upper tins
are just awash. The bridge being intended to carry infantry
in single file 6 ft. apart, how far apart should the floats be
spaced? The average weight of a fully equipped man may be
taken as 180 lb.

16. A container filled with water is shaped like a cone
with its vertex upward, its base being horizontal. Its height
is 4 ft. and the area of the base is 9 sq. ft. Calculate the total
fluid thrust due to the water on the base, and explain why
this is greater than the weight of water in the container.

17. A cylinder is partly filled with water. When a solid is
placed in it and allowed to float freely, the level of the water

rises *a* inches. When the solid is pushed down so as to be completely immersed in the water, the level of the water in the cylinder rises a further distance *b* inches. What is the specific gravity of the solid?

*18. The diagram shews the flat end of a pontoon 9 ft. wide floating in water with this end vertical, *WL* being the

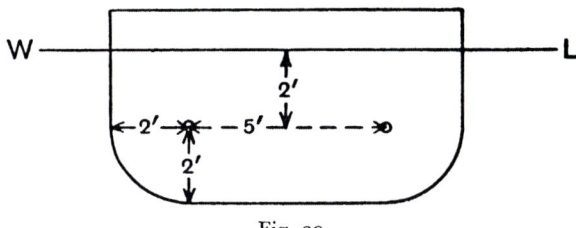

Fig. 20.

water line. Find the total water pressure on this end and the distance of the centre of pressure from the water line. The curved portions are quadrants of a circle of 2 ft. radius.

*19. Find the depth of the centre of pressure of a regular hexagon which is immersed in a uniform fluid with one side on the surface, the plane of the hexagon being vertical.

IV

STRUCTURAL PROBLEMS

1. The given observations were made in a testing machine on a bar of metal 8 in. long and $\frac{7}{8}$ in. diameter. The bar finally broke at 16 tons when the diameter at the point of fracture was $\frac{3}{4}$ in. Determine Young's modulus, the stress at yield point, the ultimate stress, the percentage extension and the percentage contraction of area.

Load in tons	0	3	6	9	10	11	12	13	14	15	16
Extension in inches	0	0·003	0·006	0·009	0·01	0·011	0·06	0·15	0·3	0·5	0·65

2. The Service D 8 signal cable consists of one strand of copper wire of diameter 0·02 in. and seven strands of steel wire, each of diameter 0·015 in. The yield point of the copper is 7 tons per sq. in., that of the steel being 15 tons per sq. in. E is 5500 tons per sq. in. for copper and 13,000 tons per sq. in. for steel. Calculate which material will reach the yield point first when the cable is given a tensile load, and if this load is 40 lb. find the stress then produced in the steel and in the copper.

3. Shew that if a wire is stretched between two points so that its sag is small compared with the span, its form will approximate to a parabola.

A telephone wire, made of bronze of density 550 lb. per cub. ft., is suspended between two points at the same level and 200 ft. apart so that the stress in the wire is 2 tons per sq. in. What is the approximate sag at the middle of the span?

4. A retaining wall is being supported by copper tie-bars of circular cross-section and 1 in. in diameter, secured to hold-fasts. A total pull of 40 tons has to be withstood, the ultimate stress of copper is 15 tons per sq. in. and a factor of

safety of 5 is considered suitable. How many of these bars must be employed?

5. The forces on the axle of a loaded railway truck are as follows:

(1) A pair of upward vertical forces each of magnitude 5 tons cutting the axle at points 4 ft. 9 in. apart.

(2) A pair of downward vertical forces each of magnitude 5 tons cutting the axle at points 9 in. outside the other pair of points.

Determine the correct diameter of the axle if the longitudinal stress is not to exceed 2·5 tons per sq. in.

6. A copper fire-box stay in a locomotive boiler is 10 in. long and 1 in. diameter. It is supporting 20 sq. in. of surface under a pressure of 200 lb. per sq. in. By how much will the stay contract in length when the pressure inside the boiler falls to 15 lb. per sq. in., the temperature remaining unaltered? E for copper is $14·5 \times 10^6$ lb. per sq. in.

7. A bridge whose clear span is 24 ft. is to carry a gun with a single axle load of 16 tons. Good quality pitch-pine balks, of cross-section 10 in. by 8 in., are available. Find how many road-bearers will be required if the governing consideration is that the maximum longitudinal stress in the material is not to exceed 1 ton per sq. in. and if the stress due to the dead loading is negligible. Take a live load factor of $\frac{3}{2}$.

8. Two plates, each $\frac{1}{2}$ in. thick, are to be connected by a double riveted lap joint, rivets of $\frac{7}{8}$ in. diameter being employed. Design the principal dimensions of the joint and calculate its efficiency. The allowable stresses are 6 tons per sq. in. under tension, 5 tons per sq. in. under shear and a bearing stress, resisting crushing, of 10 tons per sq. in.

9. A propeller shaft is 2 in. in diameter. It is running at 240 revolutions per minute and is transmitting 32 horse-

power. Calculate the torque in the shaft and the maximum shearing stress that is produced.

If the shaft is connected to the next length by flat-faced couplings which are bolted together by five bolts, each of $\frac{1}{4}$ sq. in. cross-sectional area on a pitch circle 5 in. in diameter, what is the shearing stress in the bolts?

10. A boiler, diameter 3 ft., made of $\frac{1}{4}$ in. steel plates, is subjected to internal pressure of 50 lb. per sq. in. Calculate the longitudinal and hoop stress set up in the plates.

11. A tension member of a timber frame is 3 in. wide by 4 in. deep, and to strengthen it two steel plates, each $\frac{1}{4}$ in. wide and 4 in. deep, are bolted on, one on each side of the tie. If the total tensile load is 5 tons, how much is carried by the timber? Find also the stress in the timber and in the steel.

E for steel = 30,000,000 lb. per sq. in.; E for timber = 1,500,000 lb. per sq. in.

12. A turning tool for use in a lathe may be considered as a cantilever encastred at the tool-post, and carrying a vertical load at the point due to the cutting. If such a tool is of square section, $\frac{1}{2}$ in. × $\frac{1}{2}$ in., and projects 3 in. from the toolholder, find the maximum load due to cutting that it can support if the maximum stress due to bending in the tool is not to exceed 10,000 lb. per sq. in.

If this load is applied when turning work 6 in. diameter, at 60 revolutions per minute, find the horse-power required to drive the lathe.

13. A frequent specification for good cast iron is that a bar 1 in. wide and 2 in. deep laid across a span 3 ft. wide should support a central load of 30 cwt. Find by theoretical calculation the greatest tensile stress set up in the material by this loading.

Actually cast iron will not sustain this tensile stress. Explain the apparent discrepancy.

14. A horizontal girder has one end built into a wall, the other end being unsupported and projecting 10 ft. from the wall. It carries a uniformly distributed load of ½ ton per foot run over its whole length, and also concentrated loads of 2 tons and 4 tons at its unsupported end and half-way along respectively. Draw the shearing force and bending moment diagrams for the girder.

15. A vertical column is of circular cross-section and 12 in. in diameter. It is loaded with a weight W which is large in comparison with the weight of the column and at a distance a from it. By considering this load as being equivalent to an equally centrally applied load combined with a bending moment Wa, determine the maximum value of a that will admit of no part of the column being in tension.

16. The diagram gives the result of a tensile test on a specimen of aluminium bronze 10 in. long and 0·2 sq. in. cross-section. Determine the elastic limit in tons per sq. in.,

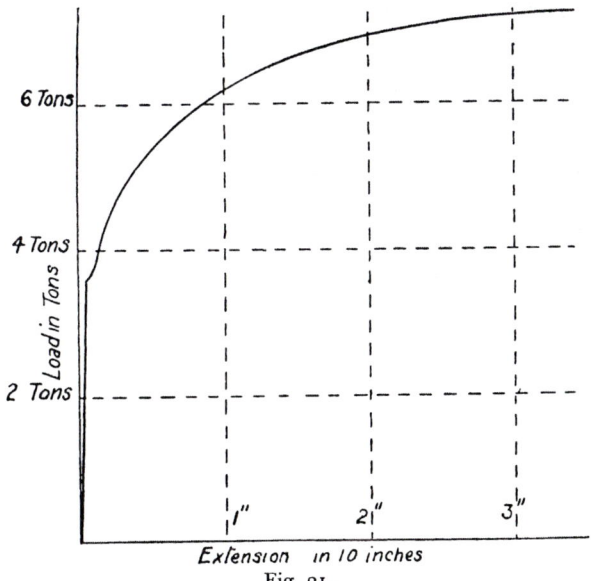

Fig. 21.

the ultimate stress, the percentage extension and the work done per cub. in. in breaking it.

17. The figure shews the framework of a small aeroplane wing. The lift forces which it is expected to sustain are

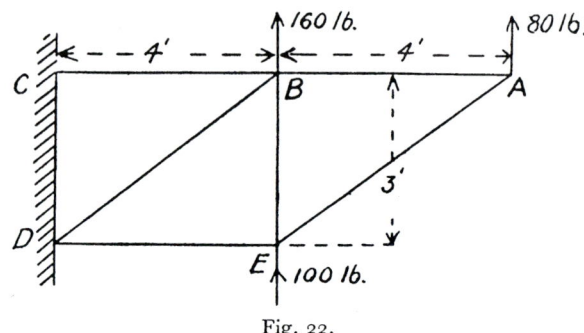

Fig. 22.

indicated. Tabulate the stresses in the various members, distinguishing those that are struts from those that are ties.

18. Shew that a 15 ft. gap can be bridged by a single 10 in. × 10 in. pitch-pine road-bearer on each side so as to carry a gun with a single axle load of 4 tons. The weight of the bridge itself can be neglected, but a live load factor of $\frac{3}{2}$ should be introduced. The maximum allowable longitudinal stress for pitch pine is 1 ton per sq. in.

19. The beam ABC, of negligible weight, is supported at A and B, AB being 15 ft. long and BC 3 ft. The middle third of the portion AB is loaded with a distributed load of 1 ton per foot run, and in addition there are two concentrated loads of 2 tons, one 3 ft. to the left of B and the other at the

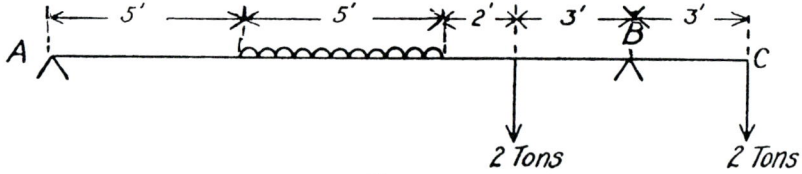

Fig. 23.

overhanging end C. Draw the diagrams of shearing force and of bending moment. At what points is the bending moment zero?

20. The cross-section of a pillar is as shewn, being formed of three similar girders connected together. Each girder is of outside dimensions 5 in. by 10 in., webs and flanges being all 1 in. thick. Find the radius of gyration of the cross-section (1) about XX', (2) about YY'.

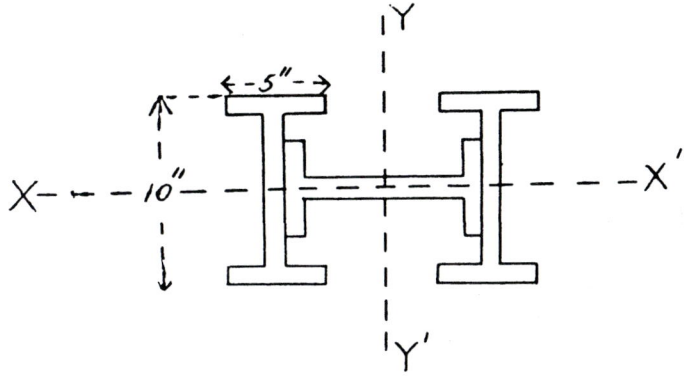

Fig. 24.

21. A suspension bridge with two cables is to be made for infantry in file across a span 160 ft. wide. The superstructure weighs 80 lb. per foot run, and infantry in file produce a moving load of 280 lb. per foot run, which may be considered as equivalent to a dead load of 420 lb. per foot. A dip of $\frac{1}{10}$ the span is arranged for. Find what size of cables should be employed. For steel cable the working rule to be used is that $9c^2 = $ load in cwt., c being the circumference of the cable in inches.

*22. In the framework indicated by Fig. 25, $ABCF$, $FCDE$ are squares and G is the middle point of FC. Assuming that the framework is pin-jointed throughout and that it supports a vertical load of 1 ton, determine the stresses in the members *AF*, *BC*, *AG*, *BG*.

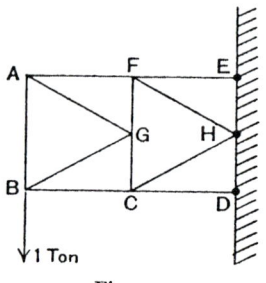

Fig. 25.

23. An iron beam is of rectangular cross-section, having a constant breadth of 4 in. but a varying depth. It is built in horizontally at one end and carries a single concentrated load of 1 ton at the free end, the total unsupported length being 8 ft. In comparison with this load the weight of the beam itself may be neglected. The maximum stress at all cross-sections is to be 6 tons per sq. in. Calculate the depth at the built-in end and draw a curve shewing how the depth varies along the beam from the built-in end to 1 ft. from the free end.

24. Construct a stress diagram for the given loaded framework. The five members *AB*, *BC*, *CD*, *CE* and *BE* are all of equal length. It is supported at *E* and at *D*. Calculate the stresses in the horizontal members and distinguish struts from ties.

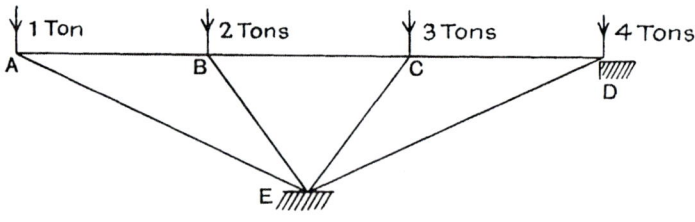

Fig. 26.

25. The figure shews a crane carrying a load of 1000 lb. The horizontal arm CE is hinged to the vertical member AF and is supported by the strut BD which is hinged at both ends. The footstep at A takes the whole weight of the structure. Find the magnitude and direction of the reactions at A, at F and at C.

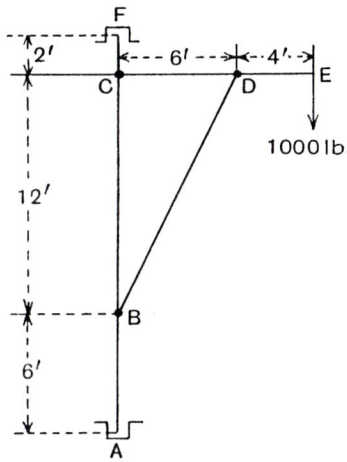

Fig. 27.

26. What force is required to punch a hole of diameter 1 in. in a copper plate $1\frac{1}{2}$ in. thick if the ultimate shearing stress of copper is 12 tons per sq. in.?

27. A knuckle joint as shewn connects two portions of a tie-rod and the pull in the rod is 4 tons. Assuming that there is no bending of the pin, find its least permissible diameter d, the allowable shearing stress being 3 tons per sq. in.

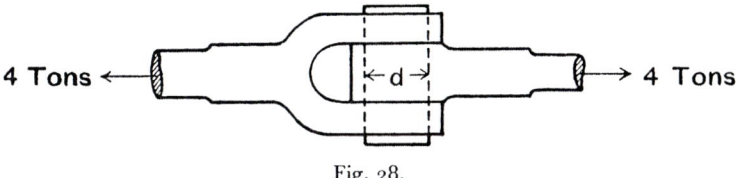

Fig. 28.

28. The side member of the chassis of a motor lorry is 6 in. deep, its cross-section being given in the figure. It can be regarded as a beam 12 ft. long, freely supported at its ends and carrying a load of 2000 lb. uniformly distributed along its length, its own weight being included in this loading. Calculate the moment of inertia about its neutral axis, and thus find what is the maximum intensity of stress in the member.

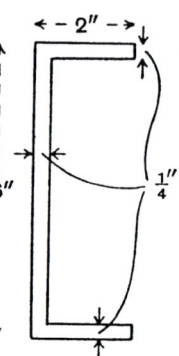

Fig. 29.

29. A uniform bar is suspended horizontally by two vertical steel wires, one at each end, and by a vertical copper wire at the middle. A load of 50 lb. is hung from the bar at a point midway between one end and the middle. Find the extra tensile stress in each wire, if each of them has a diameter of $\frac{1}{20}$ in. and if initially they are all of the same length.

Assume that the bar does not bend and that Young's modulus for steel is twice that for copper.

V

HEAT AND HEAT ENGINES

Take the weight of 1 cub. ft. of dry air at 0° C. and at a pressure of 14·7 lb. per sq. in. (76 cm. of mercury) as 0·08 lb.

1. 1 lb. of air at 27° C. and at an initial pressure of 15 lb. per sq. in. is heated at constant volume to 227° C. What is the new pressure?

It is now further heated at this new pressure, which remains constant, so as to expand to twice its original volume. Find (*a*) the temperature at the end of the expansion, (*b*) the heat that is absorbed by the air in each of the two operations, (*c*) the number of ft.-lb. of work that it will do in expanding.

2. In one method of manufacturing lead pipes the solid lead is squirted through an annular die of the required shape under pressure of a plunger working on the solid lead in a cylinder. If the pressure on this plunger is 24,000 lb. per sq. in., find the rise of temperature in the lead as it is squirted through the die. The kinetic energy of the emergent lead may be neglected. Specific gravity of lead = 11·5. Specific heat of lead = 0·03.

3. A balloon contains 10,000 cub. ft. of hydrogen when at such height that the atmospheric pressure is that of 65 cm. of mercury and when the temperature both inside and outside the envelope is − 13° C. If the deflated balloon, together with its crew and basket, weighs 650 lb., calculate whether it will rise further or sink. The density of air is 14·4 times that of hydrogen.

4. A pendulum of a clock consists of a heavy weight at the end of a light metal rod. If the complete period of the pendulum is one second at a temperature of 10° C., how many

seconds will the clock lose in 24 hours if the temperature is raised to 25° C.? Coefficient of linear expansion of rod = 0·000017 per °C.

5. An iron sphere has a diameter of 6 in. at 0° C. What will its diameter be and what will be its percentage increase of volume at 200° C.? The coefficient of linear expansion of iron is 0·000012 in °C. units.

6. A hot-water radiator allows 0·5 thermal unit radiation per square foot of surface per hour for every degree difference between the water and the air. A room is to be kept at 15° C. and in cold weather 750 thermal units are required every hour for this purpose. What radiator surface will be required if the temperature of the water in the radiator can be raised to 90° C.?

7. If the density of carbon monoxide is 0·1232 lb. per cub. ft. at 0° C. and 14·7 lb. per sq. in. pressure, find the value of the constant R in the equation $PV = RT$ in ft.-lb. sec. units.

8. 0·1 oz. of coal is burnt in a coal calorimeter which contains 7 lb. of water, the temperature of which rises from 14° C. to 19° C. With 2 grammes of coal of calorific value 9000 C.Th.U. per lb. and the same calorimeter the temperature of the water rose from 14° C. to 18° C. What was the water equivalent of the apparatus and the calorific value per lb. of the 0·1 oz. of coal?

9. An analysis of Bovey Tracey lignite shewed its combustible contents to be 66 per cent. carbon and $2\frac{3}{4}$ per cent. hydrogen. Taking the calorific value of carbon as 14,500 B.Th.U. per lb. and that of hydrogen as 61,000 B.Th.U. per lb., estimate the calorific value of this lignite.

In a test of 100 lb. of this fuel, only 1,100,000 B.Th.U. were evolved. It was noted that the fuel was very wet. Estimate how much water was contained in each lb. of lignite. The latent heat of steam in °C. units is 540.

10. A vessel contains air and mercury at a certain pressure, the volume of the air being 175 c.c. and its temperature 14° C. The vessel is heated and it is necessary to run 204 grammes of mercury out to keep the pressure the same as before. Find the rise in temperature. The density of mercury is 13·6 and its expansion may be neglected.

11. A kettle of water is raised to the boiling point from 12° C. by a gas jet in 10 min. Taking the latent heat of steam as 540, and assuming that the heat is supplied to the kettle at a constant rate, shew that there is no danger of the kettle boiling dry in an hour.

12. The heat required to change 1 lb. of water into steam at the same temperature is 970 B.Th.U., when the pressure is 15 lb. per sq. in. The volume of steam thereby produced is 26 cub. ft., the volume of the water being negligible. Shew that about 7·45 per cent. of the heat is accounted for by external work.

13. A lead bullet is fired against a steel plate with a velocity of 1000 ft. per sec. If the plate is not appreciably deformed, and if the bullet flattens out so that the kinetic energy of its fragments after impact is negligible, find what proportion of the bullet is momentarily melted. Specific heat of lead is 0·03. Melting point of lead is 620° F. Latent heat of lead is 10·5 in °F. units.

14. A chimney is 100 ft. high. The mean temperature of the gases in the chimney is 100° C. and the atmospheric temperature is 10° C. Find the chimney draught in inches of water, allowing for an increase of 4 per cent. in the density of the air in the chimney due to the presence of the products of combustion.

15. A cubical hot-water storage tank of 50 in. side is lagged with insulating material 1 in. thick. The tank is full of water at 70° C., the temperature of the outer surface of the lagging

being 20° C. The thermal conductivity of the lagging in inch Centigrade second units is 0·0003. Find approximately how much heat is conducted through each of the six sides of the tank per second and thence find how long it takes to fall 1° C. in temperature.

16. A cylinder whose volume is 25 cub. ft. contains air saturated with water-vapour at a temperature of 25° C., the pressure being 14·7 lb. per sq. in. Calculate the weight of dry air present.

17. The density of a certain brass at 0° C. is 524 lb. per cub. ft. What would be the density of the same brass if heated to 200° C.? The coefficient of linear expansion of the brass is 0·000018 per °C.

18. A kettle containing 6 lb. of water is heated on a gas ring. The initial temperature of the water is 20° C. and all the water is ultimately boiled away. The amount of gas used is 30 cub. ft. and its calorific value is 240 C.Th.U. per cub. ft. What percentage of the available heat in the gas is used in heating and evaporating the water? Latent heat of steam = 540 in °C. units.

If the gas is sold at 7d. per therm, the therm being 100,000 British Thermal Units, how much does the above gas cost?

19. A cube-shaped cold store 6 in. × 6 in. × 4 in. contains several tons of ice. The walls, roof and floor are 8 in. thick and the material of which they are made has a thermal conductivity of 2×10^{-5} in. lb.-ft. sec. units. If the average temperature of the outside of the walls is 16° C., calculate how many lb. of the ice will be melted in a week. Take the latent heat of ice as 80.

20. The stem of a mercury thermometer has a bore of 0·06 cm. diameter and the graduations are such that 5 cm. of length represents 20° C. The mean coefficient of cubical expansion of mercury relative to glass is 0·00015. Find the

weight of mercury required for the thermometer. The specific gravity of mercury at 0° C. is 13·6.

21. A gas engine working on the Otto cycle develops 15 indicated horse-power with a gas consumption of 250 standard cub. ft. per hour. The compression ratio is 3·6. Find the theoretical and the actual thermal efficiency, the calorific value of the gas being 350 C.Th.U. per standard cub. ft. Take γ as 1·4.

22. If a boiler and engine take 2 lb. of coal per indicated horse-power hour and the coal gives out 7300 C.Th.U. per lb., find the thermal efficiency of the plant.

23. A steam engine uses 100 lb. of steam per minute. The steam entering the condenser has a dryness fraction of 0·75, its pressure being 6 lb. per sq. in., and it is there condensed to water and reduced to a hot well temperature of 50° C. If the rise in temperature of the condensing water is 25° C., find how much of it is required per minute.

24. Steam at a pressure of 120 lb. per sq. in. and dryness fraction 0·8 is expanded adiabatically down to 20 lb. per sq. in. What is the dryness of the steam at the end of the expansion and what is the total heat per lb. of steam before and after the expansion?

*25. A gas-producer in which air and steam are drawn through incandescent anthracite is supplying gas consisting of hydrogen, carbon monoxide and nitrogen only. The coal contains by weight 94 per cent. carbon and 6 per cent. ash. 1 lb. of carbon produces 2450 C.Th.U. when burnt to CO and 1 lb. of hydrogen produces 34,500 C.Th.U. when burnt to form steam. Assuming no heat interchanges other than those in the chemical reactions, calculate how much steam and how much air are required per lb. of coal burnt. Air contains 23·2 per cent. oxygen and 76·8 per cent. nitrogen by weight.

26. The following results were obtained in a steam engine trial:

Dry saturated steam used per hour, 1800 lb. at 130° C.
Indicated horse-power, 120.
Cooling water through condenser per hour, 10 tons.
Inlet temperature of this water, 6° C.
Outlet temperature of this water, 45° C.
Temperature of air-pump discharge, 60° C.

Draw up an approximate heat balance for this engine and find the thermal efficiency.

27. A certain gas engine firm advertises its products as giving one brake-horse-power per penny. Criticise this statement.

The calorific value of the oil being 10,000 C.Th.U. per lb., find the hourly consumption of oil in a Diesel engine which is yielding 100 horse-power, the overall efficiency of the engine at this load being 20 per cent.

28. A simple slide valve driven by an eccentric has a travel of 4 in. The angle of advance is 30° and the lead is $\frac{1}{8}$ in. Find the outside lap, and if compression takes place when 85 per cent. of the return stroke is made, find the inside lap.

29. The following results were obtained in a trial of a gas engine: Engine speed, 256 revolutions per minute. Brake tensions, 75 lb. (taut side), 12 lb. (slack side). Brake radius, 3·5 ft. The cylinder was cooled by 16 lb. of water per minute raised through 25° C. The exhaust gases were cooled by 15 lb. of water per minute raised through 30° C. Five standard cub. ft. of gas were used per minute, the calorific value of the gas being 260 C.Th.U. per cub. ft. Draw up a heat balance giving the heat supplied and the various items of heat expenditure per minute. To what do you estimate that the "unaccounted for" heat is due?

30. Calculate the "air cycle efficiency" of the above engine whose cylinder diameter was 7 in., the length of stroke being 15 in. The clearance volume was 0·082 cub. ft. Take γ as 1·4.

The indicator diagrams in the above test averaged 1 sq. in. in area, their length being 3·75 in. The indicator spring number was 400 and there were 94 explosions per minute. Calculate the indicated horse-power and the indicated thermal efficiency. What is the mechanical efficiency as loaded?

Use any of the information given in question 29 or obtained therein.

31. A boiler working at a pressure of 100 lb. per sq. in. evaporates 2 tons of water per hour, the steam formed being dry. The temperature of the feed water is 12° C. To do this 5 cwt. of coal per hour are required, the calorific value of the coal being 8000 C.Th.U. per lb. Find the efficiency of the boiler.

32. The travel of a slide valve is $4\frac{1}{2}$ in. The angle of advance is 30°. Cut off takes place at 80 per cent. of the stroke and exhaust begins when the piston has still 2·75 per cent. of its distance to travel. Neglecting obliquity, find outside lap, inside lap and lead.

33. The following points lie on a graph giving energy dissipated per minute by an aero engine as heat and measured in horse-power against speed in revolutions per minute:

Horse-power dissipated	185	198	211	228	250
Revolutions per minute	200	400	600	800	1000

Plot a graph giving C.Th.U.s dissipated per revolution at different speeds in revolutions per minute.

34. A pumping plant uses 15 lb. of coal per hour and 10,000 gallons of water are pumped every hour up a total vertical height of 200 ft. If the efficiency of the pump itself

P E   4 is 80 per cent., calculate the brake horse-power of the pumping engine. The calorific value of the coal being 9000 C.Th.U. per lb., calculate the overall thermal efficiency of the whole plant.

35. A cylinder fitted with a piston contains 2 lb. of steam at a pressure of 100 lb. per sq. in. If the volume occupied is 5 cub. ft., what is the dryness fraction of the steam and what is its total heat?

How much heat will have to be supplied in order to evaporate all the water, the temperature and pressure remaining constant?

36. In one of the high-pressure cylinders of a liner the area of the cylinder was 2000 sq. in., the mean pressure (top and bottom) was 90 lb. per sq. in. and the mean velocity of the piston in feet per minute during the stroke was 840. What horse-power was being developed in the cylinder?

37. A man working 8 hours a day does work at the rate of one-tenth of one horse-power. A boiler has an efficiency of 70 per cent. and supplies steam to an engine whose thermal efficiency is 15 per cent. For how many days will the man have to work in order to give out as much useful work as can be obtained from one ton of coal used in the above plant? The calorific value of the coal is 9000 C.Th.U. per lb.

38. A steel scale is accurately divided when at a temperature of 15° C. When the temperature is 38° C. it is used to measure the length of a copper rod. The reading is 198·35 cm. What would be the true length of the rod at 0° C.? The coefficient of linear expansion for steel is 11×10^{-6} per ° C., and that for copper is 18×10^{-6} per ° C.

VI

MAGNETISM, STATICAL ELECTRICITY AND ELECTROLYSIS

1. A magnet completes 50 oscillations in 1 min. 40 sec. at a place where H is 0·18. What time will be required for 50 oscillations at a place where H is 0·2?

2. Three bar magnets have poles of strengths 300, 400 and 400 respectively. They are arranged so that their north poles are at the corners of an equilateral triangle of 5 in. side, their south poles being so distant that their effect may be neglected. What is the force acting on the pole of strength 300?

3. A straight bar magnet of effective length 20 cm. and pole strength 100 c.g.s. units is placed on a horizontal table. Another bar magnet of effective length 20 cm. and pole strength 60 c.g.s. units is placed on the table at right angles to the first magnet and so that its centre is on the prolongation of the axis of the first, the distance between the centres being 20 cm. Find the torque in dyne-centimetres that will be produced on the second magnet.

4. Two magnets, of pole strengths 40 and 20 and each 10 cm. long, are arranged to form two parallel sides of a square $ABCD$, the one at AB having its north pole at A and the other at CD having its north pole at C. Find the magnetic force (i) at the centre of the square and (ii) at the mid-point of the side BC.

5. A condenser of 2 microfarads' capacity is in series with one of 3 microfarads' capacity. The outer plate of the second is earthed and a potential of 1000 volts is applied to the outer plate of the first. At what potential is the inner plate of the combination and what is the charge on each plate?

*6. The insulation resistance of a 3 microfarad condenser is 100 megohms. It is charged to 100 volts and then isolated. Find the time that will elapse before the voltage across the condenser drops to 50 volts (i) approximately, by considering the average current, (ii) accurately, by considering the quantity of electricity that passes for a small potential drop and integrating up for the whole drop.

7. What is the potential of a sphere of radius 5 cm. which is given a charge of 50 units?

A concentric spherical "earthed" conductor, radius 20 cm., is now arranged around it. What is now the potential of the sphere?

8. Two rain drops, a considerable distance apart, are of radii $\frac{1}{2}$ and $\frac{1}{3}$ millimetre and are at potentials of 20 and 15 units respectively. Find what change in the potential energy due to the charge will ensue if they coalesce.

*9. A condenser of capacity C farads is charged to a certain p.d. and is then discharged through a non-inductive resistance of R ohms. Shew that the time taken from the beginning of discharge until the p.d. between the plates is half its original value is about $0.69RC$.

10. How much energy is expended in conveying a charge of 20 units from a place where the potential is 10 to another place where it is 30?

11. A charged insulated sphere, radius 10 cm., is surrounded by a concentric uncharged spherical conductor of radius 20 cm. The outer conductor is "earthed" and the inner conductor is given a charge of 100 units. Find the resultant potential of the inner sphere and the capacity of the system.

12. Shew how to combine 3 condensers each of 2 microfarads' capacity so as to produce a capacity of 3 microfarads.

*13. A condenser is charged to a definite voltage and is then connected to an electrostatic voltmeter. It is found that in 15 min. the voltage has fallen a certain amount, due to leakage through the insulation resistance of condenser and voltmeter.

The experiment is repeated twice, first with a known resistance of 5 megohms connected across the condenser in addition to the voltmeter, and secondly with an unknown resistance replacing the 5 megohms. It is found that the times for the voltage to fall through the same range as before are 4 min. and 10 min. respectively.

Calculate the value of the combined insulation resistance of the condenser and electrostatic voltmeter in parallel and that of the unknown resistance.

14. A 50 microfarad condenser is charged to a potential of 100 volts and its positive and negative plates are then respectively connected to those of an uncharged 40 micro-farad condenser. Find the potential that will exist between the plates after this has been done, and the energy that has been dissipated by doing so.

15. A condenser is connected through a galvanometer and a resistance of 1 megohm to D.C. mains of 200 volts. After a time the galvanometer reads 120 microamperes. What is the P.D. to which the condenser is charged at this instant?

*16. When a condenser of capacity C is charged from the mains through a resistance R, the current I flowing into the condenser in t seconds after connection is given by the equation

$$CR\frac{dI}{dt} + I = 0.$$

If $C = 0.0005$ and $R = 800$, find what period of time elapses before the current is reduced to one-tenth of its initial value.

17. A sphere, radius 16 cm., has a charge of 400 units. What energy must be expended to increase the charge to 800 units?

18. It is found that when a gramme of hydrogen is burnt to form water 34,000 calories of heat are given out. Calculate the minimum E.M.F. necessary to decompose water. The electrochemical equivalent of hydrogen is 10.44×10^{-6} grammes per coulomb.

19. Two Daniell cells, each of E.M.F. 1·1 volts and resistance 0·25 ohm, are arranged to drive a current through a resistance of 1·6 ohms. Estimate the consumption of zinc per hour when the cells are (i) in series, (ii) in parallel. The electrochemical equivalent of zinc is 0·00034.

20. A number of ornaments are to be plated with a total weight of 500 grammes of silver. The current is supplied at 10 volts, and the electrochemical equivalent of silver is 0·00112. Find the cost of the electric energy used at a half-penny per unit.

21. An old-fashioned type of meter consists of a zinc voltameter shunted by a resistance which is $\frac{1}{99}$ that of the voltameter. Calculate the number of coulombs of electricity passed through the main circuit for every gramme of zinc deposited in the voltameter. How many units does the deposition of one gramme represent if the voltage of the system that is being metered is 200? Electrochemical equivalent of zinc is given in question 19.

22. A square plate of 50 cm. side is to be plated with copper on both sides to a depth of 1 mm. The current available is 50 amperes. Find how long it will take. The specific gravity of copper is 8·6 and the electrochemical equivalent of copper is 0·000328.

23. In the preparation of aluminium by the electrolytic process the current through each furnace is 8000 amperes. The electrochemical equivalent of aluminium being 0·0000936, what weight of aluminium should theoretically be obtained from each furnace per hour?

The voltage of the dynamo is 4 and the actual potential required to produce the decomposition of the aluminium oxide is 2·8. What is the resistance of each furnace?

If the actual yield of aluminium is 50 per cent. of that theoretically expected, what is the cost of the electrical energy required for producing a ton of aluminium at $\frac{1}{2}d$. per unit? What bearing do these figures have on the absence of aluminium factories in England?

VII

OHM'S LAW. POWER AND ENERGY

1. The P.D. of a battery of accumulators is 96 volts when discharging at the rate of 50 amperes, but on switching over to a charging current of 40 amperes the P.D. becomes 106. Calculate the E.M.F. and the resistance of the battery.

2. The insulation resistance of a telegraph line between two stations A and B is 12,000 ohms, and between A and an intermediate station C it is 20,000 ohms. What is the insulation resistance of the section CB?

3. Conductors having resistances of 20, 30, 40 and 40 ohms respectively are arranged to form the sides of a square $ABCDA$. The points B and D are joined by a conductor of such a resistance that a current of 0·25 ampere flows from B to D. If the P.D. between A and C is 200 volts, what current flows in the other conductors and what is the P.D. between B and D?

4. A galvanometer gives a reading of 1 division for every $\dfrac{1}{10,000}$ ampere passing through it. Its resistance is 4 ohms. Calculate the resistance of a shunt that would make the instrument into an ammeter reading 5 divisions per ampere.

5. The specific resistance of copper is $\frac{2}{3} \times 10^{-6}$ ohm per inch cube and its specific gravity is 8·8. The specific resistance of aluminium is 1×10^{-6} ohm per inch cube and its specific gravity is 2·6. Compare the diameter and the weight of two transmission lines made of copper and of aluminium respectively, whose resistances are to be equal.

*6. A pair of D.C. mains is working at 200 volts P.D. A voltmeter of 1000 ohms' resistance reads 90 volts between the positive main and earth and 50 volts between the negative main and earth. What are the insulation resistances between positive and earth and negative and earth respectively?

7. A Leclanché cell A has an E.M.F. of 1·4 volts and a resistance of 3 ohms. A Daniell cell B has an E.M.F. of 1·1 volts and a resistance of 2 ohms. They are connected in parallel across a resistance R. Find the value of R such that the current supplied by A shall be double that supplied by B. Find also the total current that will flow through R.

8. A telegraph route between A and C consists of an underground cable 12 miles long from A to B, in series with an overhead wire 8 miles long from B to C. The end at C is disconnected and insulated and a battery of 100 volts is connected in series with a milliammeter between the end A and earth. With the overhead wire and the cable joined at B this ammeter reads 0·0015 ampere, but when the connecting link is removed the reading is only 0·0004 ampere. Find the insulation resistances of the cable and of the overhead wire.

9. A dynamo is feeding a pair of mains, each 1000 yd. long. At 400 yd. from the dynamo 120 amperes are taken off, as are 80 amperes at the end of the mains. If the pressure at the supply end of the mains is 200 volts, find that at the two points of delivery, the power lost in the mains and the efficiency of transmission. The resistance of each cable is 0·05 ohm per 1000 yd.

*10. Two similar mains, 400 yards long, each have a resistance of $\dfrac{1}{10,000}$ ohm per yard. A uniform load is taken from the positive main throughout its length at the rate of 1 ampere per yard and returned in like manner through the negative main. One end of the pair is kept at 205 volts P.D. and the other end at 200 volts P.D. Find the current flowing in at each end and the minimum voltage between the mains.

*11. A constant P.D. of 100 volts is applied to the terminals of a copper coil whose temperature is 0° C. when its resistance is 50 ohms. The temperature coefficient of the coil is 0·004 and it loses heat at the rate of 4 watts for every degree rise of temperature above 0° C. Find the final steady temperature that will be reached.

If the specific heat and the weight of the coil are known, write down in the form of an integral an equation whereby the time for a given rise of temperature to be reached could be found.

*12. The graph connecting the current through a lamp and the P.D. between its terminals is in general a straight line, and for such a lamp when the current is 0·2 ampere the P.D. is 1·8 volts, and when the current is 0·25 ampere the P.D. is 2·3 volts. Current is supplied to the lamp by a battery consisting of two small dry cells A and B, each of 4 ohms' resistance and E.M.F. 1·4 volts, connected in parallel, and used in series with a larger cell C of resistance 1 ohm and E.M.F. 1·4 volts. What current does the lamp take? Is it a metal filament or a carbon filament lamp?

13. A rail motor-car weighing 20 tons is driven by a petrol engine which drives a dynamo, the latter supplying current to motors which are geared to the driving axles. The normal running speed is 60 ft. per sec. when the tractive resistance is 16 lb. per ton. The current supplied to the motors is 180 amperes at a P.D. of 220 volts, and when thus working the engine consumes every hour 50 lb. of petrol of calorific value 20,000 B.Th.U. per lb. Find the efficiency of the motors and gearing and the overall efficiency of the whole arrangement when running at the above speed on the level.

14. A telegraph wire AB is 10 miles long and has a resistance of 5 ohms per mile. It develops a partial earth at an unknown point C. The operator at A signals to the operator at B to insulate his end and the resistance between A and earth is then found to be 110 ohms. When the end at B is fully earthed the resistance measured between A and earth is

46 ohms. Find the distance in miles from B to C and the resistance of the partial earth at C.

15. The resistance of a galvanometer is 80 ohms, and it is in series with another resistance of 100 ohms and a battery of constant E.M.F. whose resistance is 20 ohms. The galvanometer is then shunted with a resistance whose value is such that its deflection is reduced to one-half of its former value. Calculate the resistance of this shunt.

*16. A tramcar taking a current of 60 amperes runs along a line, the trolley wire voltage of which is maintained at the same potential at the two ends of a section $\frac{1}{2}$ mile long, the resistance of the section being $\frac{1}{2}$ ohm. What is the average power loss in the trolley wire?

17. The resistance in a motor starter is in circuit for 5 sec. and during this time loss of heat by radiation and conduction may be neglected. The material of which the resistance is made has specific gravity of 9, specific heat of 0·1, and specific resistance of 20 microhms per cm. cube. If the temperature rise in the 5 sec. is not to exceed 100° C., what is the greatest permissible current density in amperes per sq. cm.? Temperature effect on the resistance may be neglected.

18. A voltmeter of 40,000 ohms' resistance and a condenser are connected in series and a P.D. of 250 volts is applied. The reading of the voltmeter rises to 3 volts. What is the insulation resistance of the condenser?

19. A small wireless set requires for its high-tension supply a current of 20 milliamperes at 120 volts. Dry batteries for the purpose cost 10s. each and are worn out after delivering 4 ampere hours. If the set is required for 400 hours per year, how many dry batteries are required per annum? Compare the cost of battery use with that of an "eliminator" working from the mains. It may be assumed to have an overall efficiency of 20 per cent. and the cost of electric power is 5d.

per unit. The cost of interest and depreciation for the "eliminator" may be taken as 1s. per month.

20. Electric power is to be taken to a place 1 mile (2 miles of cable) from a generator and the current density in the cables is to be 1000 amperes per sq. in. of copper. At what voltage must the electric power be generated if a total loss of power of 10 per cent. in the cables is to be allowed? The specific resistance of copper is $\frac{2}{3} \times 10^{-6}$ ohm per inch cube.

21. A dry battery has its terminals joined to a voltmeter whose resistance is 600 ohms and the reading is 2 volts. On inserting a second and similar voltmeter in series with the first the reading of each becomes 1·05 volts. What is the E.M.F. of the battery and its resistance?

22. A coil of wire whose resistance is 8 ohms is immersed in a tank of oil, and a steady P.D. of 120 volts is applied to its terminals. It is known that radiation losses are 25 watts for every 1° C. rise above 15° C., which is the temperature of the air. Find the final steady temperature of the tank (i) if the wire is made of manganin of zero temperature coefficient, (ii) if the wire is made of copper of temperature coefficient 0·004.

*23. A wire rectangle ACFD has its side AC twice as long as its side AD. The mid-points of the two longer sides are joined by a wire so that two squares are formed side by side. If the resistance of each side of these two squares is 1 ohm, find the equivalent resistance between A and F.

24. A potentiometer wire has a length of 1 metre. When connected up in the usual way a balance is obtained with the slider at 72·4 cm. when using a standard cell of E.M.F. 1·433 volts, and with the slider at 54·3 cm. when using another cell. Find the E.M.F. of this other cell.

The resistance of the standard cell is 1900 ohms and a current of 10^{-6} ampere can be detected by the galvanometer, whose resistance is 100 ohms. Through what distance in millimetres on either side of the balance point can the slider be moved without the error being visible on the galvanometer?

25. Twelve cells are arranged in each of the following ways:

(a) Three rows in parallel each with four cells in series.

(b) Twelve cells in series.

(c) Twelve cells in parallel.

Each cell has a resistance of 0·4 ohm and an E.M.F. of 1·2 volts, and the external resistance which is in series with the whole battery is 4 ohms. Calculate the current through the 4 ohms' resistance in each case.

*26. Two leads, each of resistance 0·3 ohm and length 300 ft., are used for lighting a row of lamps which may be considered as constituting a uniformly distributed load of 0·1 ampere per ft. for the last 100 ft. of the pair of leads. If a P.D. of 100 volts is supplied at the end of the leads remote from the lamps, find the P.D. that will then exist between the other end, and the total loss of power in the leads.

27. Fuse wires of the same material are supposed to have current-carrying capacity proportional to (diameter)$^{\frac{3}{2}}$. Give the reasoning and assumptions upon which this is based.

Use the above to compare (a) the diameters, (b) the lengths, of the filaments of a 200 volt 16 c.p. lamp and a 50 volt 32 c.p. lamp. Assume both filaments to be of the same material and heated to the same temperature when in use.

28. It is required to deliver 2000 kilowatts at a P.D. of 500 volts, at a point 10 miles from a dynamo, with a total loss in transmission of 10 per cent. of the power delivered. What will be the cost of the copper conductors (double) if the copper costs £35 a ton? The density of copper is $\frac{1}{3}$ lb. per cub. in. and the specific resistance of copper is $\frac{2}{3}$ microhm per inch cube.

29. A thermos flask containing 1 litre of water at 90° C. is found to cool at the rate of 2° C. per hour. If it contains a coil of 700 ohms' resistance, what P.D. would have to be maintained between the terminals of this coil to maintain its temperature at 90° C.?

*30. Three batteries, A, B and C, are joined up with their positive terminals all connected together and their negative terminals all connected together. A has an E.M.F. of 10 volts and an internal resistance of 10 ohms. B has an E.M.F. of 12 volts and an internal resistance of 12 ohms. C has an E.M.F. of 15 volts and an internal resistance of 15 ohms. Find the currents passing through each battery and the common P.D. between the positive and negative terminal.

*31. A Wheatstone bridge consists of a resistance AB of 50 ohms, a resistance BC of 10 ohms, a resistance AD of 40 ohms, while the resistance DC is unknown. There is a cell giving a P.D. of 3 volts between A and C, A being positive. The bridge is slightly out of balance, so that a minute current flows through the galvanometer from B to D, this current being negligible compared with the currents in the arms of the bridge. From the galvanometer reading it is known that there is a potential of 20 millivolts between its terminals. Find the P.D. between B and C and between D and C, and thus determine the value of the unknown resistance CD.

32. Sketch a simple type of switchboard for a shunt dynamo containing positive and negative busbars, double pole switch, cut outs on both sides, field regulating resistance, voltmeter and ammeter.

Two shunt dynamos, A and B, are connected in parallel and together are supplying 800 amperes. The E.M.F. of A is 213 volts, that of B is 210 volts. What current is each supplying, and what is the P.D. between the brushes? The armature resistance of each is 0·03 ohm.

33. Compare the cost of heating using (a) electricity at 1d. per unit, (b) gas at 5s. per 1000 cubic feet. The declared calorific value of the gas is 540 B.Th.U. per cubic foot.

34. A pair of copper mains have a total weight of 2000 kg. and a total resistance of $\frac{1}{2}$ ohm. With a P.D. of 200 volts at one end they are suddenly short-circuited at the other. Find the resulting rise of temperature in 20 sec. The specific heat of copper is 0·095.

35. A "half-watt" lamp, supplied with a current of 0·1 ampere at a pressure of 200 volts, illuminates the screen of a photometer at a distance of 50 in. to the same extent as a standard 16 candle power lamp at a distance of 32 in. Find the candle power of the given lamp and state whether it is really a "half-watt" lamp. How do these lamps differ from ordinary low efficiency metal filament lamps?

36. In order to standardise an ammeter it was connected in series with a storage battery and a potentiometer wire 100 cm. long and of exactly 5 ohms' resistance. The positive pole of a Clark cell was connected to the same end of the potentiometer wire as the positive pole of the storage battery. A galvanometer and protecting resistance were in series with the Clark cell, this branch circuit being completed by a wandering lead on the potentiometer wire. It was found that no deflection occurred in the galvanometer when contact was made with a point on the wire 29·5 cm. from the positive end. Sketch the arrangement and find the percentage error of the ammeter if its reading was 1 ampere. E.M.F. of Clark's cell is 1·434 volts.

37. A pair of uniform feeders run along a street 1000 yd. long, the resistance of each being 0·15 ohm. A constant potential difference of 300 volts is maintained between them at the supply end; currents of 30, 40 and 50 amperes are tapped off at distances from the supply end of 400, 600 and 1000 yd. respectively. Find the difference of potential that will exist

between the mains at each of these points. Find also the efficiency of power transference of the whole arrangement.

If the specific resistance of copper is $\frac{2}{3}$ microhm per inch cube, what is the diameter of the feeders?

38. Two batteries of unequal E.M.F.s, and each of internal resistance 20 ohms, are joined up in series so as to oppose each other, and are connected in series with a resistance of 50 ohms and a galvanometer whose resistance is 60 ohms. They are then put in series, so that their E.M.F.s act in the same direction, and by increasing the series resistance to 200 ohms, and by shunting the galvanometer with 30 ohms, the reading of the galvanometer is brought back to its former value. The E.M.F. of the battery having the greater voltage being 16 volts, find that of the smaller.

39. The field coil of a dynamo has a resistance of 50 ohms at 10° C., which is the temperature of its surroundings. The coil loses heat by radiation and conduction at the rate of 10 watts for every degree Centigrade that its temperature is above that of its surroundings. Its resistance rises 0·2 ohm for every degree rise in temperature. It is found that its steady temperature is 60° C. Find what current it is taking, and at what P.D.

* 40. A and B are the positive and negative earth lamps on the switchboard for mains of P.D. 100 volts. C is a lamp connected to the mains and it is controlled by a switch S, which lies between C and the positive main. An earth, reducing the insulation resistance to 100 ohms, occurs between C and S. If the resistances of A, B and C are 100 ohms each, find the currents through A, B and C (i) when S is out, (ii) when S is in.

41. The specific resistance of copper is 1·6 microhms per centimetre cube, whereas that of aluminium is 2·9 microhms per centimetre cube. Compare the sectional area and also

the weight of two cables of equal length and resistance made of copper and aluminium respectively. The density of copper is 8·8 and that of aluminium is 2·6.

42. Heat is supplied electrically at the rate of 6 watts to a body of weight 40 grammes, and the temperature increases with the time as shewn in the table. Plot a graph shewing the specific heat at different temperatures. It may be assumed that all the energy supplied is used in raising the temperature of the body.

Time in minutes	0	0·5	1	1·5	2	2·5	3
Temperature in ° C.	20	27·5	35·3	43·3	51·5	59·8	68·3

VIII

ELECTROMAGNETISM, INDUCTION AND ELECTRODYNAMICS

1. A tangent galvanometer consists of a coil of 25 wires 30 cm. in diameter placed in the magnetic meridian. The current in the coil causes the compass needle at the centre of the coil to deflect through 30°. The value of the horizontal intensity of the earth's magnetism at the place being 0·18, find the current in amperes.

2. A wrought-iron anchor ring is of 24 cm. mean diameter and 6 sq. cm. cross-sectional area and has a flux of 60,000 lines of force produced by a coil of 400 turns carrying 1·2 amperes. Find the permeability of the iron at this flux density.

3. A closed soft iron annular ring of 25 cm. mean diameter and 4 sq. cm. cross-section is wound uniformly with 160 turns of insulated wire. The relation between the induction B and the permeability μ of the iron is as follows:

B	10,200	12,000	13,700
μ	2,000	1,500	1,000

What current must pass through the wire to produce a total flux of 50,000 lines? How much would the current have to be increased to produce the same flux if a single air-gap 1 mm. wide was made in the ring?

*4. A current of 5 amperes is flowing in a wire which is bent in the form of a square of 10 cm. side. By considering the current in a small element of the wire, find the magnetic force at the centre of the square and shew the direction of the flux with reference to the current.

*5. Prove that the force at a point distant r cm. from a wire carrying a current I amperes is equal to $2I/10r$ dynes.

Two long straight wires are parallel to one another and 10 cm. apart, and are carrying currents of 50 and 100 amperes. Find the force per centimetre between the wires.

6. A bar magnet is 12 cm. effective length and of pole strength 20. A circular coil of radius 5 cm. and having 10 turns of wire is placed with its centre distant 8 cm. from the centre of the magnet and on the line through the centre at right angles to the magnet's length. Explain in what plane this coil should be placed in order that the magnetic field at its centre may be zero when a current is passed, and find the current in amperes that is then required.

7. In a horse-shoe electromagnet the flux density at the ends of the magnet amounts to 10,000 lines of force per sq. cm. If the area of each end is 8 sq. cm., what is the maximum load in kilogrammes that the magnet is capable of holding up?

8. A circular coil of wire has a mean diameter of 20 cm. and consists of 200 turns. A current of 5 amperes flows through it and the north pole of a long magnet of pole strength 10 units is placed at the centre of the coil. Find the magnitude and direction of the force exerted on the pole by the current.

*9. Calculate (i) approximately and (ii) exactly, the magnetic force at the centre of a cardboard tube, 25 cm. long and 5 cm. radius, when it is wound with 200 turns and is carrying a current of 20 amperes.

10. The exciting current of a dynamo's field winding produces a flux of 4 million lines per pole and there are 8000 turns per pole. If the current is reduced to zero in 0·1 sec., what is the mean E.M.F. that is induced in the winding?

11. A horizontal conductor 5 ft. long is falling vertically with constant velocity 10 ft. per sec. through a horizontal magnetic field whose strength is 1000 lines of force per sq. cm. What E.M.F. is induced in the conductor?

12. A coil carrying a current has an induction of 25 henries when the current in it is 40 milliamperes. If the current is switched off in $\frac{1}{20}$ second, what voltage will be induced across the coil?

13. The field coils of a dynamo have a total of 2000 turns and produce a flux of 4 million lines of force when carrying 2·5 amperes. Assuming that the flux is proportional to the current, what is the coefficient of self-inductance of the field circuit?

14. A magnetic circuit consists of a wrought-iron anchor ring working at high permeability, and in it is an air-gap 10 sq. cm. in area and 1 cm. long. If it is wound with 600 turns, calculate its approximate self-induction in henries. Justify any assumption you make.

15. The vertical component of the earth's magnetic field in England is 0·47 c.g.s. units. A millivoltmeter is connected between the insulated rails of a railway track, gauge 4 ft. 9 in., upon which an express train is travelling. The resistance of the voltmeter is 10 ohms, that of the rails and axles being negligibly small, and the voltmeter reads 2 volts. Estimate the speed of the train in miles per hour and find the watts that are absorbed by this arrangement.

16. A motor armature has 360 conductors spaced round on a cylinder of diameter 20 cm., the axial length of each conductor being 24 cm. The poles give a flux density that may be considered as of uniform value of 8000 lines per sq. cm. over two-thirds of the circumference and zero elsewhere. A current of 12 amperes is flowing in all the conductors. Find the force in lb. wt. that is produced on a single conductor lying in the field and hence find the torque in lb.-ft. that is acting on the armature.

17. The coil of a moving-coil ammeter of the usual type with a permanent magnet, has 20 turns and is rectangular in

shape, the length parallel to the axis of suspension being 2 cm. and the width 1·2 cm. The control spring exerts a torque of 0·4 gm. cm. per radian of deflexion of the coil. When a current of $\frac{1}{20}$ ampere flows through the coil it deflects through 60°. Find the strength of the magnetic field in the gap.

18. A tangent galvanometer consists of a number of turns in series, wound on a circle of radius 10 cm., and when a current of 7 amperes flows through the coil a field of strength 11 dynes is produced at the centre of the coil. Calculate the number of the turns.

19. A coil is to be wound to a depth of $\frac{1}{2}$ in. on a wooden bobbin 1 in. in diameter and 2 in. long, the wire being made of copper of specific resistance $\frac{2}{3}$ microhm per inch cube. The coil is to produce 10,000 ampere-turns when a P.D. of 100 volts is applied. Find the diameter of the wire required if the wire is wound tightly and the thickness of the silk insulation may be neglected.

20. An airship 50 ft. in diameter has a coil of wire of 100 turns wound round it. It is turning in a circle of $\frac{1}{4}$ mile radius at 30 miles per hour. If H at the place is 0·18 gauss, find the maximum E.M.F. generated in the coil.

21. A coil has a self-inductance of 12 henries. A current of 5 amperes is produced in the coil in $\frac{1}{2}$ sec. What E.M.F. will be produced in the coil, and in what direction?

IX

DIRECT CURRENT MACHINES AND SIMPLE ALTERNATING CURRENT PROBLEMS

1. A small shunt dynamo supplies power to a series motor which is situated at a distance, and is connected to the dynamo by mains whose total resistance is 0·5 ohm. The P.D. at the terminals of the dynamo is 200 volts, the resistances of its field and armature being 100 ohms and 0·2 ohm respectively. The dynamo is supplying 40 amperes to the motor, and the frictional, eddy currents and hysteresis losses of the dynamo are then 500 watts. Calculate (1) the E.M.F. of the dynamo, (2) its overall efficiency, (3) the back E.M.F. of the motor if its resistance is 0·3 ohm, (4) its overall efficiency if its brake horse-power is 8·5.

2. A shunt dynamo is delivering 20 kilowatts at 200 volts. Its armature resistance is 0·05 ohm and its shunt resistance is 100 ohms. The losses due to friction, hysteresis and eddy currents are 1000 watts. Find its overall efficiency and the brake horse-power of the engine driving it.

3. A circular copper disc has a radius of 30 cm. and is mounted on a shaft through its centre and perpendicular to its plane, which is rotating at 300 revolutions per minute in a uniform magnetic field parallel to the shaft. Brushes are placed on the rim of the disc and upon the axle and a P.D. of 2 volts is found to be maintained between these brushes. What is the strength of the magnetic field?

4. A series motor having a resistance of 1 ohm takes 40 amperes from 200 volt mains when running at 480 revolutions per minute. What extra resistance must be placed in series if the speed is to be reduced to 330 revolutions per minute

with the current unchanged at 40 amperes? Would you expect any change in the torque of the motor?

*5. A dynamo giving a constant P.D. is connected by leads having a resistance of 0·2 ohm to a battery whose E.M.F. is 120 volts and whose resistance is 0·1 ohm. To the battery is connected a circuit whose resistance varies between infinity and 2·5 ohms. Find the dynamo P.D. so that the charging current through the cells at no load may be equal to the current output from the cells at full load, and find the value of this current.

6. The external characteristic of a shunt dynamo is such that its open-circuit P.D. is 200 volts, and its P.D. with 50 amperes is 195 volts, the intermediate points being linear. It is in parallel with a battery of 100 cells each of which on discharge gives 1·97 volts E.M.F. The resistance of the battery is 0·02 ohm. Find how a current to the external circuit of 120 amperes is shared and the terminal P.D. that will then exist.

7. A shunt dynamo has the following open-circuit characteristic at a certain speed:

E.M.F. (volts)	400	415	435
Field current (amperes)	1·5	1·8	2·0

When run at this speed and delivering 128·3 amperes to the external circuit the armature current was 130 amperes. Taking the resistance of the armature as 0·12 ohm and neglecting armature reaction, find the resistance of the shunt.

8. A tram-car is fitted with a single series motor whose resistance is 1 ohm, which is running off 600 volt mains. When this motor is in series with an additional 5 ohms the car takes a current of 50 amperes and travels at 10 miles per hour. The extra resistance is gradually cut out and when the

speed becomes steady the motor is again taking 50 amperes. What is the new speed of the tram-car?

9. A tram-car weighing 10 tons is travelling at 20 ft. per sec. up an incline of 1 in 20, the resistance due to track and air resistance being 24 lb. per ton. If the combined efficiency of the motor and gearing is 75 per cent., what current is being taken from the trolley wire, which is at a potential of 800 volts above that of the earth?

10. In a test of a series motor it was connected to 100 volt mains, a current of 20 amperes being taken, and the speed of the motor then being 1000 revolutions per minute. Its power was absorbed by a band brake applied to its flywheel, the difference in tension between the two sides of the belt being found to be 20 lb. The circumference of the flywheel was $3\frac{1}{2}$ ft. Find the motor's overall efficiency.

If the mains are altered to 145 volts, and the current taken is the same, determine the motor's new speed. Its resistance is $\frac{1}{2}$ ohm.

11. In a Hopkinson test on two similar machines coupled together, the energy supplied from the mains was 40 amperes at 220 volts. The armature resistance of each machine was 0·05 ohm. The motor field current was 6 amperes, that of the dynamo being 7 amperes and the motor armature current was 200 amperes. Draw a diagram of the arrangement and find the efficiency of each machine.

12. A four-pole generator with lap-wound armature has a flux of $1\cdot5 \times 10^6$ lines per pole and is driven at 600 revolutions per minute, the P.D. between the brushes then being 150 volts. Assuming the armature drop as 5 per cent. of the P.D., estimate the number of armature conductors.

13. A shunt motor worked off 200 volt mains takes a current of 22 amperes. The resistances of armature and field circuits are 0·2 ohm and 100 ohms respectively. The torque

transmitted to the shaft is 32·5 lb.-ft. when running at 630 revolutions per minute. Determine the overall efficiency and estimate the various losses as far as possible.

14. A series motor used to drive a ventilating fan has two poles and two field coils. Its lowest speed is 300 revolutions per minute when the field coils, each of resistance 0·2 ohm, are in series with each other and the motor is taking 5 amperes from 100 volt mains. The torque on the armature is then 9 lb.-ft. In order to increase the speed the field coils are connected in parallel with each other and the motor now takes 8 amperes. Calculate the new speed and the new torque, making the assumption that the flux is proportional to the field current. The resistance of the armature is 0·4 ohm.

15. A motor is wave wound with four poles and 320 armature conductors. It is running at 420 revolutions per minute. The armature current is 60 amperes and a voltmeter across the brushes reads 102. The armature resistance is 0·1 ohm. Calculate the back E.M.F. and the total flux produced by each pole.

What current is flowing in each of the conductors?

16. Why is a shunt dynamo always used to charge accumulators?

A dynamo of E.M.F. 100 volts and armature resistance 0·4 ohm is connected in parallel with 50 accumulators each of E.M.F. 2·1 volts and internal resistance 0·01 ohm. They are together supplying a current of 40 amperes. How is this current shared between dynamo and battery and what is the P.D. between the brushes?

17. What A.C. voltage is required to send a current of 20 amperes through a coil whose resistance is 2 ohms and whose inductance is 0·007 henry at a frequency of 50 periods a second?

18. The strength of a direct current varies as shewn in the figure. Find the steady current that would have the same heating effect as the given variable current.

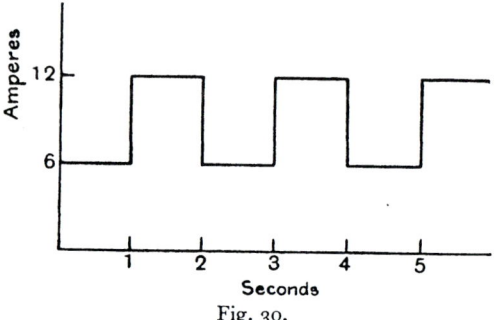

Fig. 30.

19. A choking coil of resistance 4 ohms and inductance 0·5 henry is in series with a condenser of capacity of 12 microfarads. An A.C. voltage of 120 volts is applied and the frequency is adjusted to resonance. What voltage will then be set up (a) across the choking coil, (b) across the condenser?

20. An alternator gives a current, the wave form of each half alternation being as shewn in the diagram. The maximum value of the current during each half alternation being 100 amperes, find graphically the approximate R.M.S. value of the current.

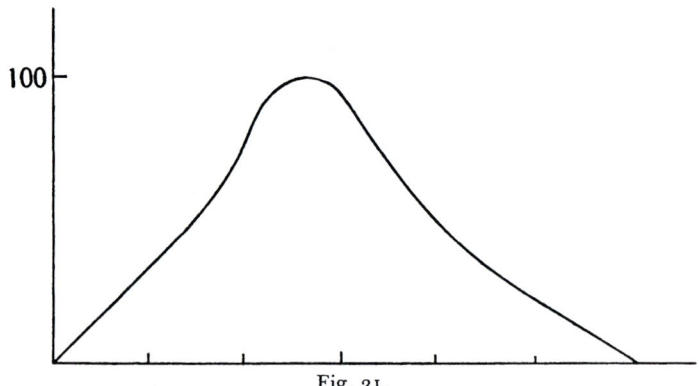

Fig. 31.

21. A circuit containing 1·7 ohms, 10 millihenries and 200 microfarads all in series is connected to A.C. mains of 120 volts at 100 cycles. Find the magnitude and phase angle of the current and the power absorbed.

22. A transformer has a normal primary voltage of 800, and on open circuit it takes a current of 0·25 ampere at a power factor of 0·75. When the secondary is short-circuited and a reduced voltage of 50 volts is applied to the primary, it takes 10 amperes at a power factor of 0·4. Calculate the efficiency when delivering the full output of 8 kilovolt amperes at unity power factor.

23. A dynamo is supplying power at 50 periods and 200 volts. The load takes 20 amperes and absorbs 3 kilowatts. Find its resistance and its self-induction. There is no capacity.

24. A 2000/200 volt, 12 kilowatt transformer absorbs 300 watts at no load, and the resistances of the primary and secondary windings are 7 ohms and 0·06 ohm respectively. It runs for 9 hours at no load, for 5 hours with a load of 25 amperes, and for 10 hours with a load of 60 amperes, the power factor being unity throughout. What is the approximate all-day efficiency of the transformer?

25. A motor giving 60 brake horse-power with an efficiency of 90 per cent. and a power factor of 0·88 (lagging) is run on 500 volt A.C. mains in parallel with a motor giving 37·5 brake horse-power with an efficiency of 84 per cent. and a power factor of 0·6 (leading). Find the resultant current taken from the mains and the joint power factor.

26. A choking coil when connected across direct current mains, the P.D. between which is 15 volts, allows a current of 4 amperes to pass, but when connected to A.C. mains of 100 volts P.D. and frequency 50 cycles, the current is found to be 16 amperes. Find the power taken from the A.C. mains, the power factor and the inductance of the choking coil.

27. A transformer has a ratio of 8 to 1, the higher pressure being 10,000 volts. The resistances of the high and low pressure windings are 4 ohms and 0·05 ohm respectively. The power absorbed at no load is 12 kilowatts. What is the efficiency of the transformer when it is delivering 400 amperes on the low pressure side to a load whose power factor is 0·8?

28. Find the virtual value of a current which rises uniformly from zero to 40 amperes in $\frac{2}{5}$ the half period, remains at that value for $\frac{2}{5}$ the half period, and falls uniformly to zero in the remaining $\frac{1}{5}$.

29. A transformer has a full load output of 20 kilowatts and a full load loss of 1000 watts. This loss is made up of 400 watts core loss and 600 watts I^2R loss. Find its all-day efficiency if working for 5 hours at full load, for 5 hours at half load and running light for the remaining 14 hours.

30. A choking coil whose resistance is negligibly small is placed in series with a bank of lamps, the two being placed across A.C. mains. The volts across the lamps are found to be 160, and across the coil they are 120. What is the approximate voltage across the mains?

31. An alternating P.D. of 100 volts at a frequency of 50 is maintained across a circuit which has a constant induction of 0·3 henry and a variable resistance. Find the value of the resistance for which the power consumed in the circuit is a maximum. What is then the power?

32. An alternating current P.D. of 250 volts is maintained across a circuit consisting of 5 ohms' resistance in series with a coil which has a resistance of 8 ohms and a reactance of 6 ohms. A 10 ohms' resistance is connected as a shunt across the terminals of the coil. Calculate the P.D. across the coil and the current that is taken from the mains by the system.

X

GEOMETRICAL AND MECHANICAL DRAWING

1. The given circle is the plan of a cone standing on its base 2 in. in diameter, its height being 2½ in. It is cut by a vertical plane *AB* as shewn, the smaller portion being removed. Draw the elevation of the cone. Draw also the true shape of the section. *XY* is the ground line.

X _____ **Y**

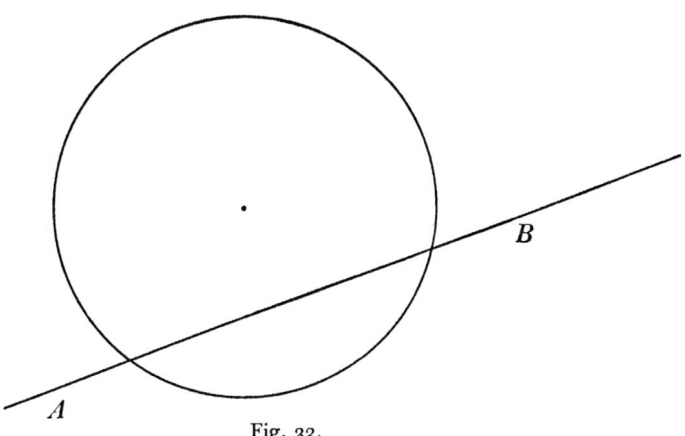

Fig. 32.

2. A cone whose height is 2·5 in. and whose base is 2 in. diameter is originally standing with its base on the horizontal plane. The top part is cut off by a plane which makes 30° with the horizontal and which bisects the axis of the cone. Draw a plan of the top portion resting with its cut face on the horizontal plane.

3. Draw the plan of a hexagonal pyramid lying on one of its triangular faces and cut by a horizontal plane *AB* as shewn on the given elevation, the upper portion being removed.

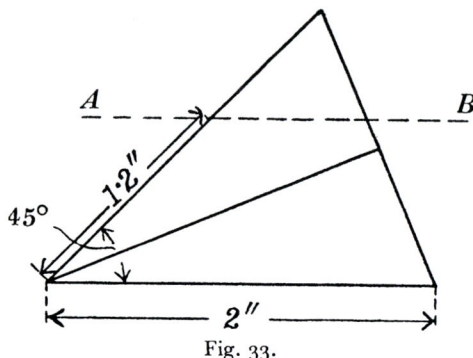

A -------------→------------- -------- *B*

45°

2″

2″

Fig. 33.

4. A cylinder whose length is 4 in. and whose diameter is 3 in. is lying on a horizontal plane, the line of contact between the curved surface of the cylinder and the plane making 30° with the horizontal ground line. Draw its plan and elevation.

5. A right pyramid has a square base of 2 in. side and its height is 4 in. It stands on the horizontal plane with its axis vertical and with two of its sides at 30° to the ground line *XY*. The upper part of the pyramid is cut off by a section plane perpendicular to the vertical plane and inclined at 45° to the horizontal and passing through a point on the axis 2 in. from the base. Draw the plan and elevation of the portion of the pyramid that is left. Draw also the true shape of the section.

6. The views given (Fig. 34) shew an outside elevation and plan of one end of an air-pump link composed of a split gun-metal bearing held together by link rods. Draw half full size (*a*) an outside end elevation looking in the direction of the arrow *Z*, (*b*) a plan view across the section *XX* looking in the direction of the arrows *YY*.

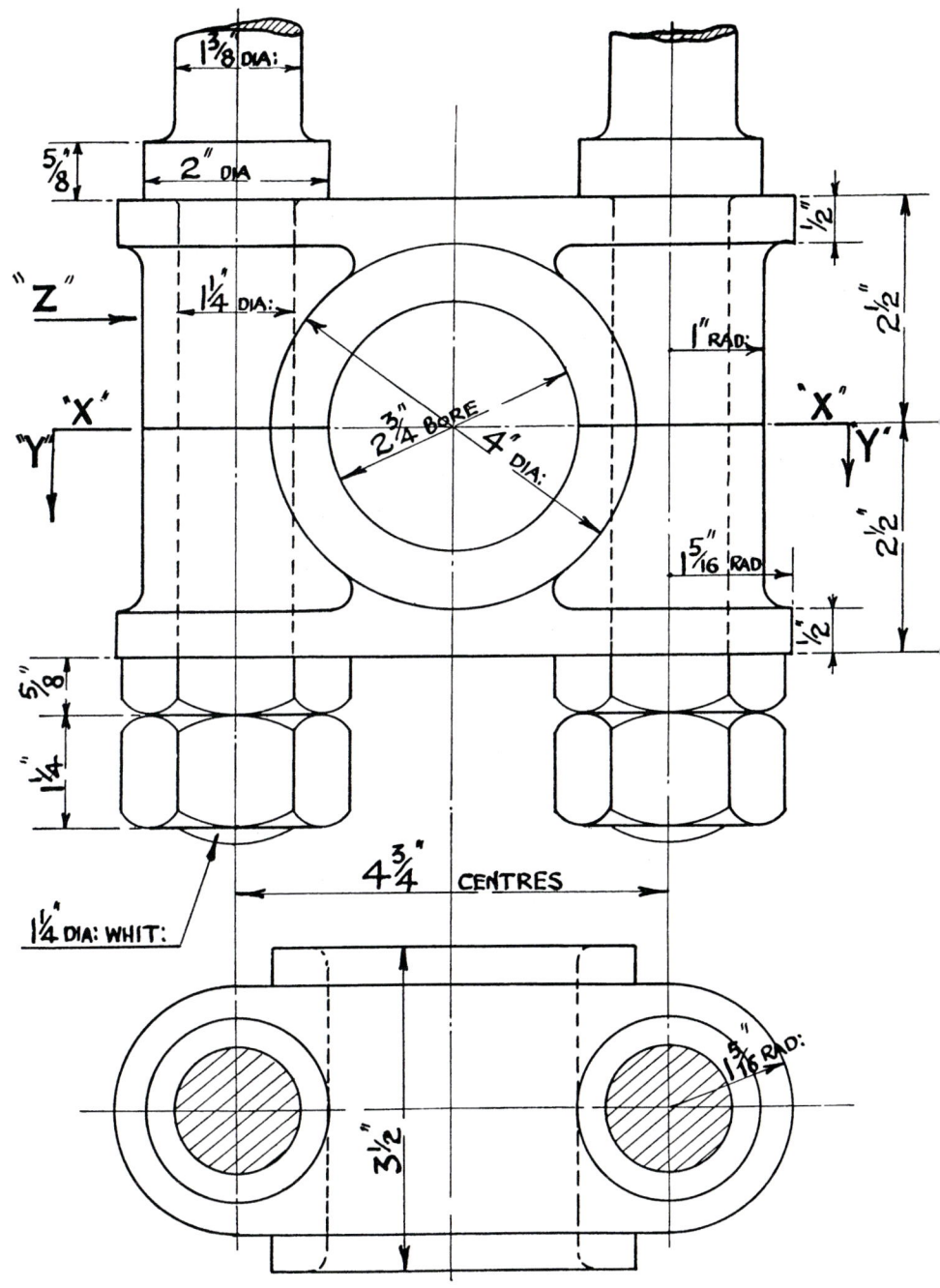

Fig. 34.

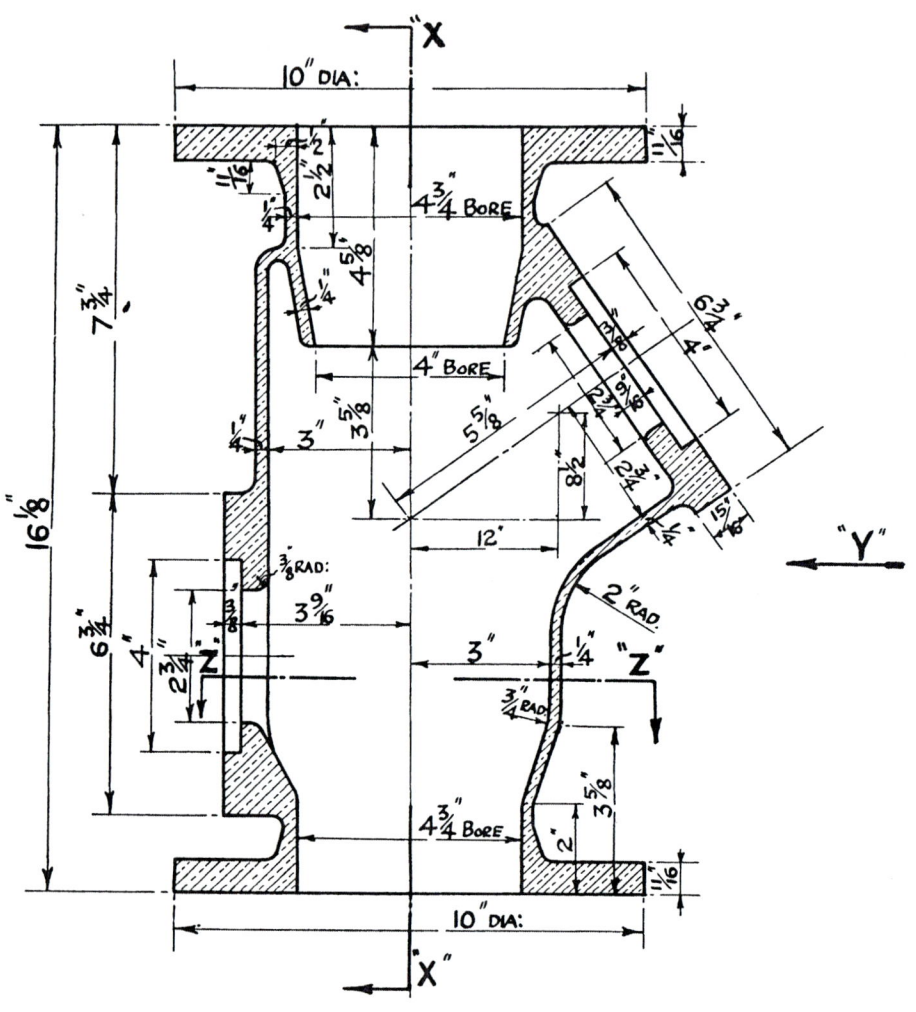

Fig. 35 a.

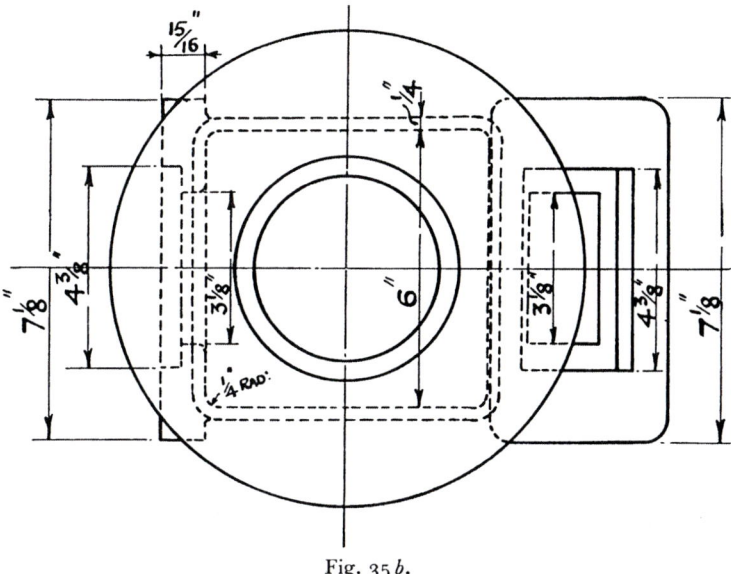

Fig. 35 b.

7. The drawings represent a sectional elevation and plan of a gun-metal casting for the body of a sight-flow indicator. Draw to a scale of half full size:

(1) A section on the plane "*XX*".

(2) An outside view looking in the direction of the arrow "*Y*".

(3) A sectional plan on the plane "*ZZ*".

(4) An outside view corresponding to the sectional elevation shewn.

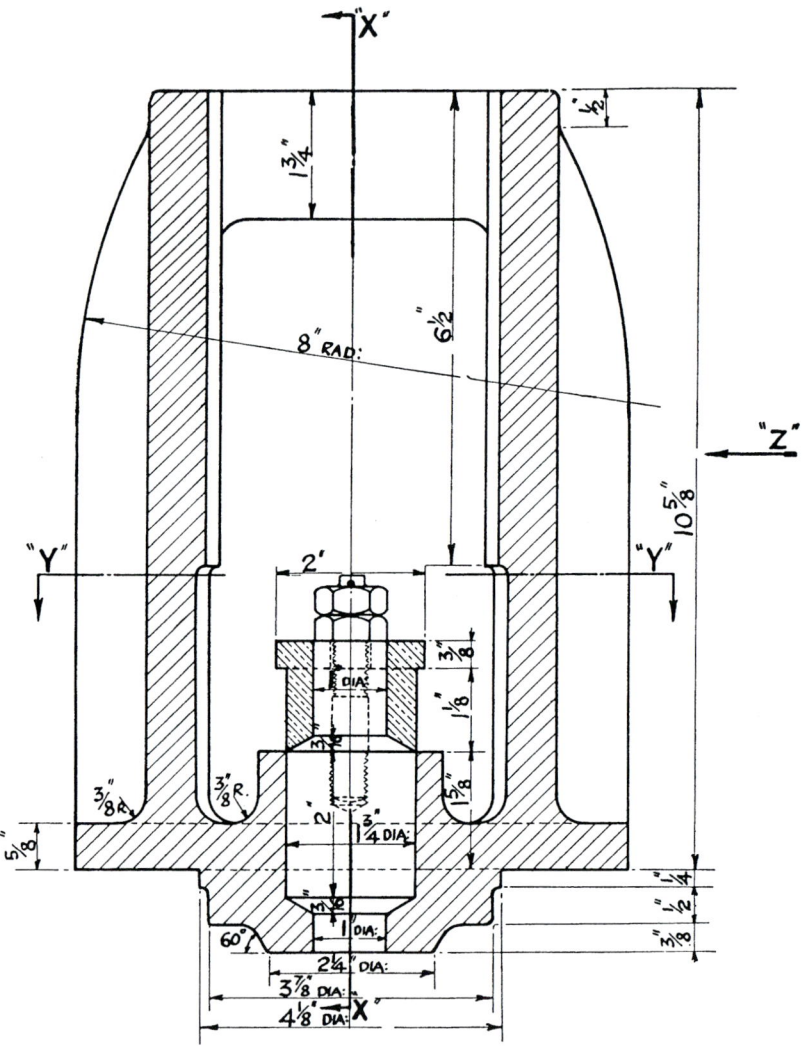

Fig. 36 a.

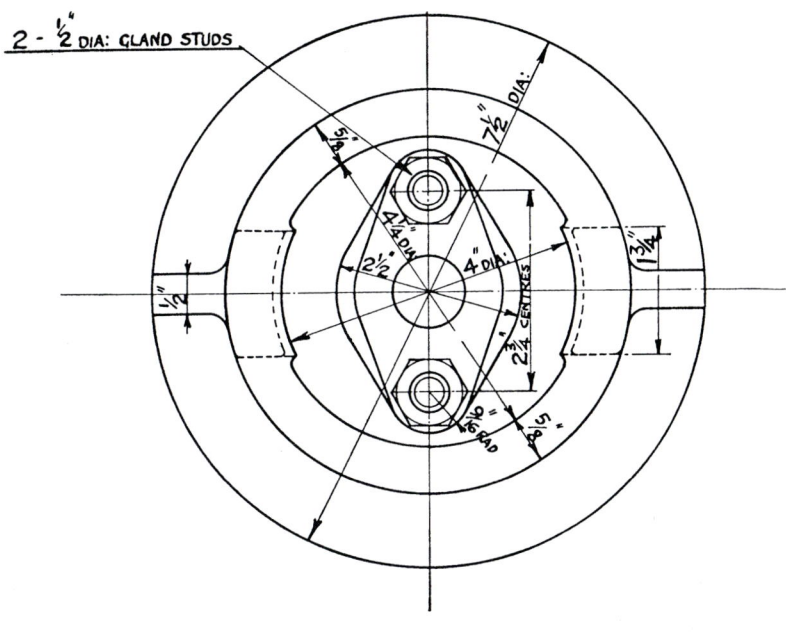

2 - ½" DIA: GLAND STUDS

Fig. 36 b.

8. The drawings represent a sectional elevation and plan of the cast-iron combined cover and piston rod guide for a small steam cylinder. Draw to a scale of half full size:

(1) A section on the plane "XX".

(2) An outside view looking in the direction of the arrow "Z".

(3) A sectional plan on the plane "YY".

(4) An outside view corresponding to the given section.

9. The drawing shews a sectional elevation of a right angle stop valve body. Draw a view looking in the direction of the arrow X, the left-hand portion shewing an outside view, and the right-hand portion shewing a section on the line AA.

Dimensions not given can be assumed.

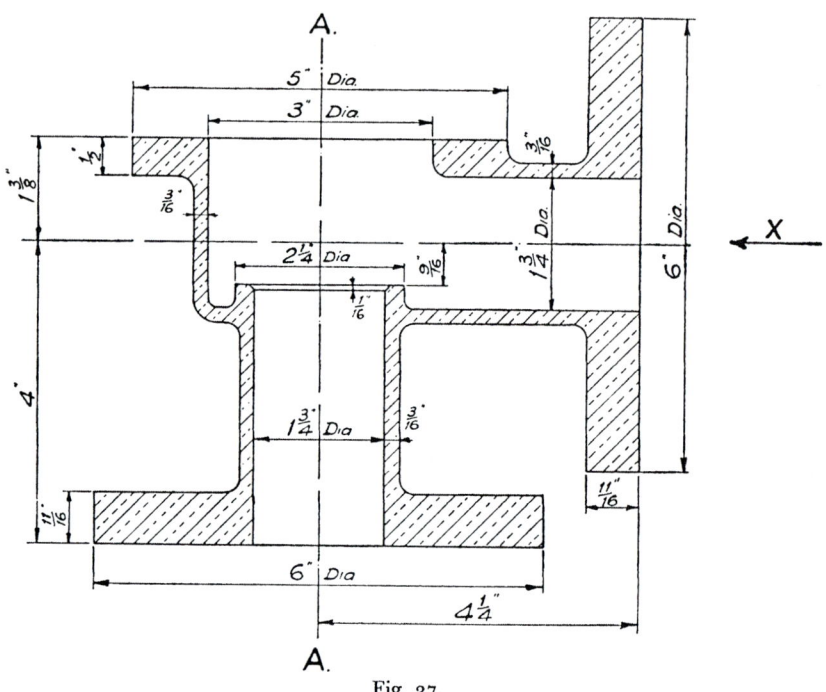

Fig. 37.

XI

ANSWERS AND SOLUTIONS

I. STATICS

1. $2\sqrt{2}$ lb.; 9·9 ft. 2. 1·44 cwt.

4. V.R. 384. Overhauling possible after 3·75 tons.

5. 119 lb. 6. 1156 lb.

7. 0·23. 8. 3 ft.

10. 13° 12′. AG nearly 60 cm.

11. $\tan^{-1} \dfrac{a+b+h}{b} \tan\theta.$

12. Approach; $\mu = 0·75$. 13. 13 lb.; 13·225 lb.

14. 1 in.; 9 cwt. 15. 7.

16. 14·67 ft. per min.; 825 ft. per min.

17. AB and AC 10·14 tons. AD 6·96 tons.

18. AB, AC, AD, tension $\sqrt{\tfrac{3}{2}}\,W$; BC, CD, DB, $\dfrac{W}{\sqrt{6}}$ compression.

19. 5573 lb.; 17·37 ft.

20. 600 lb. in each; $600\sqrt{5}$ lb.

21. 1·56 tons; 1·19 tons; approx. 2 tons.

22. $VR = 3$. It would run "chock-a-block". 120 lb.; 36 lb.; 48 lb.

23. 48·5 tons at $27\tfrac{1}{2}°$ with vertical.

24. 0·77 lb.; 4·5 lb.

26. $\dfrac{Wl}{\sqrt{9l^2-3a^2}}; \dfrac{a}{\sqrt{3l^2-a^2}}.$ 27. 5·77 kg.

28. About 47°. 29. 5376 ft.

II. DYNAMICS

1. 8 m.p.h.; S.W. 2. 4·37 m.p.h.; 12·77 m.p.h.

3. 1·43 hours; 14° 48′ N. of E. 4. 10 sec.; 90 ft.

7. 30 ft. per sec.; 18 ft. per sec.; 144 r.p.m.

8. 1445 ft.

9. 3560 ft. per sec. Sound arrives last.

11. 5·75 ft.

12. 33 ft. per sec.; 8·5 ft. per sec. per sec.; 81 ft.

13. 16 ft.; 138·6 ft.; 1·25 sec. 14. 3·92 ft.

15. 4 ft. per sec. per sec.; 3·46 sec.

16. 4·375x; 875 ft.-lb.   17. $55\rho^3 - 48\pi\rho^2 - 50\pi^3 = 0$.

18. 7·85 sec.; 2·95 rev.   19. 163·2 ft.; 4·12 sec.

20. 32 m.p.h.

21. 0·205; 80·1 H.P. It just can.

22. 8·53 sec.; 133 ft.

23. 357 H.P.   24. 1340 ft. per sec.

25. 38 ft.   26. 200 ft.

27. 10·7 ft. per sec.; 73 lb.

28. $3\frac{1}{8}$ ft. per sec.; 0·925 in. each.

29. 3790 ft. Nearly 16.

30. 0·41 ft. per sec. per sec.; 14·2 m.p.h.

31. 1·6 ft. per sec. per sec.

32. 3·2 ft. per sec.; 1·78 ft. per sec.; $5\frac{1}{3}$ lb.; 2·36 sec.

33. 1·775 m.p.h.; 9·8 lb.

34. 1,750,000 lb.-ft.; 425,000 lb.; 86,400 H.P.

35. 12·6 ft.; 78 ft. per sec.

36. 0·028 ft. per sec. per sec.; 0·57 H.P.

37. 22·56 ft. per sec.; 118 lb.

38. 1·06. No tendency to move.

39. 2 ft.; 12 ft. per sec. per sec.

40. 1·3465 kg.   41. 37,360 ft.-tons; 208 tons-ft.

42. 19 sec.   43. 114 m.p.h.

45. $93\frac{1}{3}$ sec.; 6440 ft.   46. $6\frac{6}{7}$ ft. per sec.

47. 1·75 $\sqrt{10}$ radians per sec.   48. 2450 ft.

49. 68° 50′; 27·8 lb.; 24·34 ft.-lb.; 15·07 ft.-lb.

50. 7° 40′; 8° 25′.   51. $3m$; $m\left(\dfrac{3l-a}{l-a}\right)$.

52. 3·17 sec.   53. 0·718 sec.

54. $\frac{3}{8}$ in.   55. Loses 2·4 min.

57. 29·6 ft. per sec.   58. 65 lb.-ft.

59. $\displaystyle\int_{v=0}^{v=V} \dfrac{dv}{g+Kv^2}$.   60. 88·8 per cent.

61. 34·4 m.p.h.   62. 39·3 m.p.h.; 4608 ft.

63. 1080 gm. cm.^{2} 64. 90 in.^{4}

65. 247 lb.-ft.^{2} 67. 19° 10′.

68. 44 ft. per sec.   69. No change in time.

71. 5·58 ft.   72. 9 ft. per sec.

III. HYDROSTATICS

1. $11\cdot7$ tons.   2. $8\cdot77$ cub. in.
3. 29,800 lb.
4. $6\frac{1}{2}$ in. more forward; $6\frac{1}{2}$ in. less aft.
5. $2°\,52'$.   6. $13\cdot9$ grammes.
8. 1706 lb.; 853 lb.
9. $27\cdot7$ ft. per sec.; $22\cdot6$ ft. per sec.
10. 4500 lb.; 18,000 lb.-ft.; $4\cdot82$ ft. from toe.
11. 2620 lb.; $1\frac{2}{7}$ ft. from bottom; 10,000 lb.; 6000 lb.
12. Approx. $24°$.   13. $4\cdot83$ ft.; $2\cdot23°$.
14. $3\cdot3$ ft.   15. Every 9 ft.
16. 2250 lb.   17. $\dfrac{a}{a+b}$.
18. 4621 lb.; $2\cdot595$ ft.
19. $\frac{23}{18}$ depth of geometrical centre.

IV. STRUCTURAL PROBLEMS

1. 13,300 tons per sq. in.; $19\cdot1$ tons per sq. in.; $26\cdot6$ tons per sq. in.; $26\cdot53$ per cent.
2. The steel. 15 tons per sq. in.; $6\cdot35$ tons per sq. in.
3. $4\cdot27$ in.   4. 17.
5. $5\cdot7$ in.   6. $0\cdot00325$ in.
7. 14.   8. Pitch $= 2\cdot875$ in. Efficiency $69\cdot5°/_0$.
9. 701 lb.-ft.; 5360 lb. per sq. in.; 2690 lb. per sq. in.
10. 3600 lb. per sq. in.; 1800 lb. per sq. in.
11. $1\cdot15$ tons; 215 lb. per sq. in.; 4300 lb. per sq. in.
12. $8\cdot7$ lb.; $0\cdot0248$ h.p.
13. $20\cdot25$ tons per sq. in. Cast iron is not elastic.
15. $1\frac{1}{2}$ in.
16. 18 tons per sq. in.; 36 tons per sq. in.; $34°/_0$; $9\cdot4$ inch tons.
17. Struts: BA, 107 lb.; BC, 558 lb.; BE, 180 lb. Ties: DE, 107 lb., AE, 133 lb., BD, 566 lb.
19. $1\cdot5$ ft. to left of B and at ends.
20. $K_x = 3\cdot08$ in.; $K_y = 5\cdot05$ in.   21. $7\frac{1}{4}$ in. circumference.
22. Struts: BC and AG, 1 and $1\cdot12$ tons. Ties: AF, BG and AB, 1, $1\cdot12$ and $\frac{1}{2}$ ton.
23. $4\cdot9$ in.

24. Ties: *AB*, 1·73 tons; *BC*, 2·88 tons; *CD*, 1·15 tons. Struts: *AE*, 2 tons; *BE*, 2·3 tons; *CE* 3·45 tons; *DE*, 1·33 tons.
25. 1120 lb.; 500 lb. and 675 lb.
26. 56·4 tons.   27. 0·925 in.
28. 11·75 inch units; 9250 lb. per sq. in.
29. Steel, 16,550 and 3850 lb. per sq. in. Copper, 5100 lb. per sq. in.

V. HEAT AND HEAT ENGINES

1. 25 lb. per sq. in.; 727° C.; 34 and 120 C.Th.U.; 47,000 ft.-lb.
2. 114° C.
3. Net buoyancy is 25 lb., i.e. it will rise.
4. Nearly 11 sec.   5. 6·0144 in.; 0·72 per cent.
6. 20 sq. ft.   7. 62·8.
8. 2·92 lb.; 7936 C.Th.U. per lb.
9. 11,247 B.Th.U. per lb.; 0·46 lb. of water.
10. 24·6° C.
13. 33 per cent.   14. 0·315 in.
15. 37·5 C.Th.U.; 20 sec.   16. 1·79 lb.
17. 518·4 lb.   18. 51·7 per cent.; 0·91 penny.
19. 610 lb.   20. 64 grammes.
21. Theoretical 40 per cent. Actual 24·3 per cent.
22. 9·65 per cent.   23. 1766 lb.
24. 0·94; 564 C.Th.U.; 610 C.Th.U.
25. 0·6 steam; 3·1 lb. air.   26. 16 per cent.
27. $70\frac{5}{7}$ lb.   28. 0·87 in.; 0·55 in.
29. Gas 1300; b.h.p. 253; jacket 400; exhaust 450; unaccounted 197.
30. 47·8 per cent.; 14·5 h.p.; 26 per cent.; 74·5 per cent.
31. 65 per cent.   32. 0·92 in.; 0·46 in.; 0·22 in.
34. 12·65 h.p.; 10·6 per cent.
35. 0·562; 892 C.Th.U.; 432 C.Th.U.
36. 4580 h.p.   37. 1870 days.
38. 198·265 cm.

VI. MAGNETISM, STATICAL ELECTRICITY AND ELECTROLYSIS

1. 94·8 sec. 2. 1290 dynes.
3. 310 dyne-centimetres. 4. 0·565 dyne; 2·2 dynes.
5. 400 volts; 1200 micro-coulombs.
6. 200 sec.; 208 sec. 7. 10 E.S.U.'s; 7·5 E.S.U.'s.
8. 0·436 erg. 10. 400.
11. Potential, 5 E.S.U.'s; capacity, 20.
13. 13·75 megohms; 27·5 megohms.
14. 55·5 volts; 0·117 joule. 15. 80 volts.
16. 0·92 sec. 17. 15,000 E.S.U.'s.
18. 1·5 volts.
19. 2·56 grammes; 0·78 gramme.
20. 0·62 penny.
21. $2·94 \times 10^5$ coulombs; 16·35 units.
22. 72·8 hours.
23. 2·7 kg.; 0·00015 ohm; £12. 10s. per ton.

VII. OHM'S LAW. POWER AND ENERGY

1. 101·55 volts; $\frac{1}{9}$ ohm. 2. 30,000 ohms.
3. Current in AB 4·15 amperes; in AD 2·375 amperes. 12 volts.
4. 0·002 ohm.
5. $\dfrac{D_{AL}}{D_{CU}} = 1·225$. $\dfrac{W_{AL}}{W_{CU}} = 0·444$.
6. 1200 ohms; 667 ohms. 7. 4·22 ohms; 0·225 ampere.
8. 250,000 ohms; 90,900 ohms.
9. 192 volts; 187·2 volts; 1984 watts; 95·04 per cent.
10. 262·5 amperes; 137·5 amperes; 198·1 volts.
11. 42·5° C. Time $= \displaystyle\int_{\theta=0}^{\theta=\theta} \dfrac{\rho W d\theta}{4\theta - \dfrac{200}{1 - 0·004\theta}}$ where W is the water

equivalent of the coil.
12. 0·2304 ampere. Metal filament.
13. 66 per cent.; 8·9 per cent. 14. 4 miles; 80 ohms.
15. 48 ohms. 16. 300 watts.
17. 1940 amperes. 18. 3·29 megohms.

19. 2 dry batteries. Battery, 20*s*.; eliminator, 14*s*.
20. 845 volts.   21. 2·21 volts; 63 ohms.
22. 87° C.; 73·75° C.   23. 1·4 ohms.
24. 1·076 E.M.F.; 1·01 mm.
25. 1·06 amperes; 1·64 amperes; 0·298 ampere.
26. 95 volts; 46$\frac{2}{3}$ watts.   27. $\frac{D_1}{D_2} = \frac{1}{4}$; $\frac{L_1}{L_2} = 2$.
28. £4460.   29. 40·5 volts.
30. $I_A = 0·2$ ampere; $I_B = 0$; $I_C = -0·2$ ampere; 12 volts.
31. 0·5 volt; 0·48 volt; 7·62 ohms.
32. 450 amperes; 350 amperes; 199·5 volts.
33. 2·24 : 1.   34. 2° C.
35. 39 c.p. Yes approximately.
36. 0·972 ampere; 2·8 per cent.
37. 285·6 volts; 280·2 volts; 274·2 volts; 93 per cent.; 0·44 in.
38. 11 volts.   39. 2·89 amperes; 173 volts.
40. $\frac{3}{5}$ ampere; $\frac{2}{3}$ ampere; $\frac{1}{5}$ ampere. $\frac{1}{3}$ ampere; $\frac{2}{3}$ ampere; 1 ampere.
41. Area, 16 : 29; weight, 1·87 : 1.

VIII. ELECTROMAGNETISM, INDUCTION AND ELECTRODYNAMICS

1. 0·1 ampere (approx.).   2. 1250.
3. 3·61 amperes; 9·8 amperes.   4. 0·566 dyne.
5. 0·0314 gramme.
6. 0·192 ampere or 3·89 amperes.
7. 65 kg.   8. 200π dynes.
9. 201 dynes; 187 dynes.   10. 3200 volts.
11. 0·461 volt.   12. 20 volts.
13. 32 henries.   14. 0·045 henry.
15. 66 m.p.h.; 4×10^{-7} watts.   16. 0·52 lb.; 41 lb.-ft.
17. 1732 lines per sq. cm.   18. 25 turns.
19. 0·02 in.   20. 0·01 volt.
21. 120 volts.

IX. DIRECT CURRENT MACHINES AND SIMPLE ALTERNATING CURRENT PROBLEMS

1. 208·4 volts; 86·5 per cent.; 168 volts; 88 per cent.
2. 92·1 per cent.; 29·1 H.P.   3. 1420 lines per sq. cm.
4. 1·25 ohms. No.

5. 124·75 volts; 15·8 amperes.
6. 45 amperes; 75 amperes; 195·5 volts.
7. 231 ohms.   8. 18⅓ m.p.h.
9. 61·5 amperes.   10. 79 per cent.; 1500 r.p.m.
11. Motor, 90 per cent.; dynamo, 89·4 per cent.
12. 1050.   13. Nearly 81 per cent.
14. 371 r.p.m.; 11·5 lb.-ft.
15. 96 volts; 2·14 × 10^6 lines; 15 amperes.
16. Battery, 23⅓ amperes; dynamo, 16⅔ amperes; 93⅓ volts.
17. 59·5 volts.   18. Nearly 9·5 amperes.
19. 6120 volts; 6120 volts.   20. 51 amperes.
21. 50 amperes leading 45°; 4250 watts.
22. 95·5 per cent.   23. 7·5 ohms; 0·021 henry.
24. 92·1 per cent.   25. 170 amperes; 0·98.
26. 960 watts; 0·6; nearly 0·016 henry.
27. 93 per cent.   28. 31 amperes.
29. 91·9 per cent.   30. 200 volts.
31. 94·2 ohms; 53 watts.   32. 130 volts; 24·5 amperes.

X. GEOMETRICAL AND MECHANICAL DRAWING

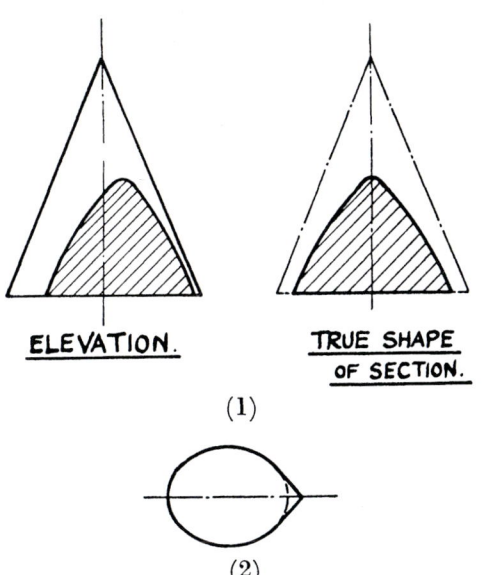

ELEVATION.   TRUE SHAPE
                  OF SECTION.

(1)

(2)

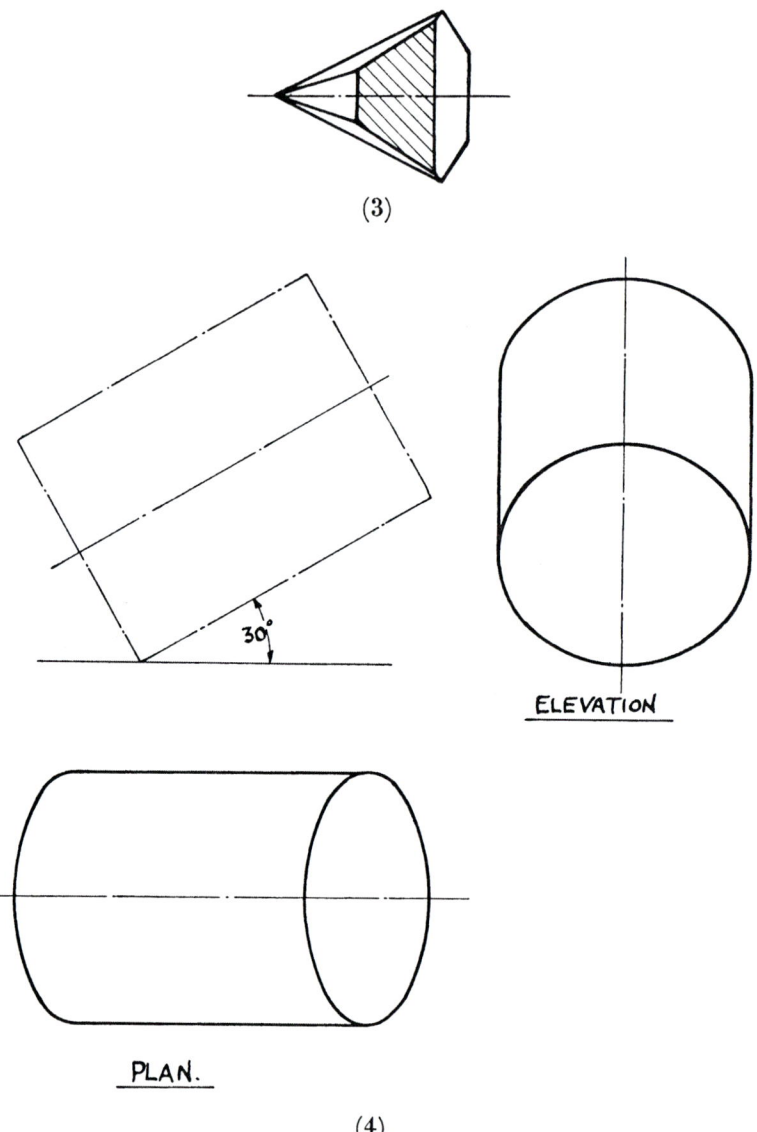

(3)

30°

ELEVATION

PLAN.

(4)

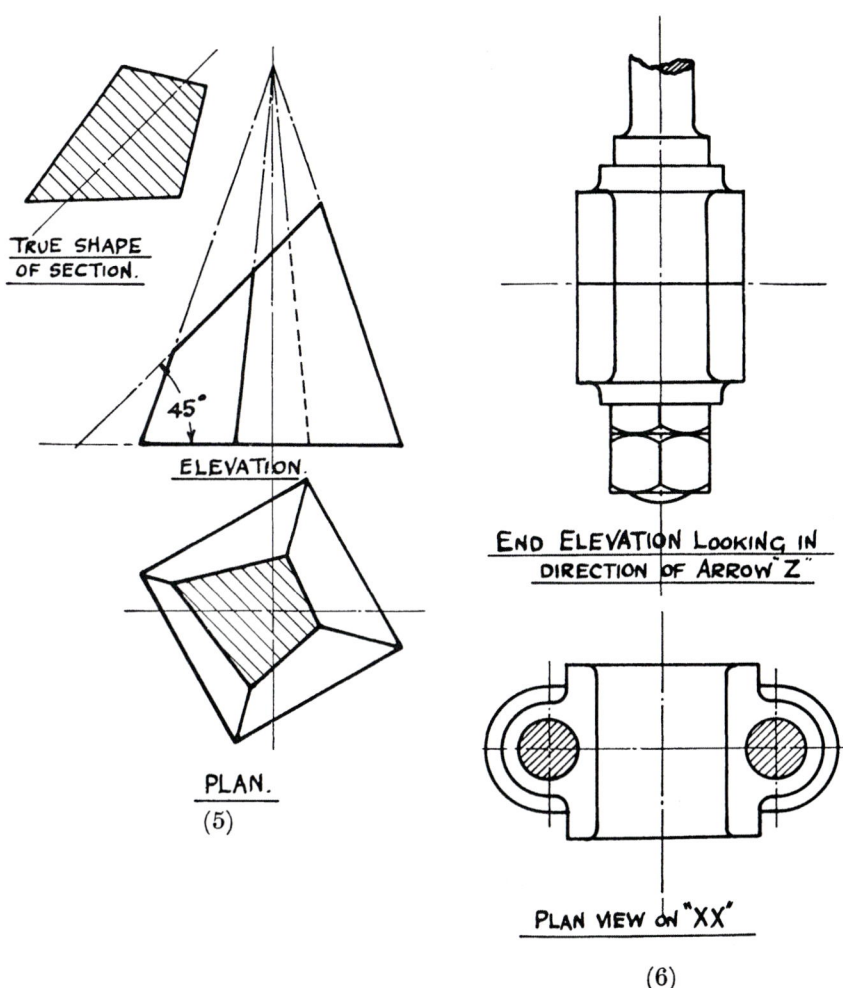

TRUE SHAPE
OF SECTION.

45°

ELEVATION.

PLAN.

(5)

END ELEVATION LOOKING IN
DIRECTION OF ARROW "Z"

PLAN VIEW ON "XX"

(6)

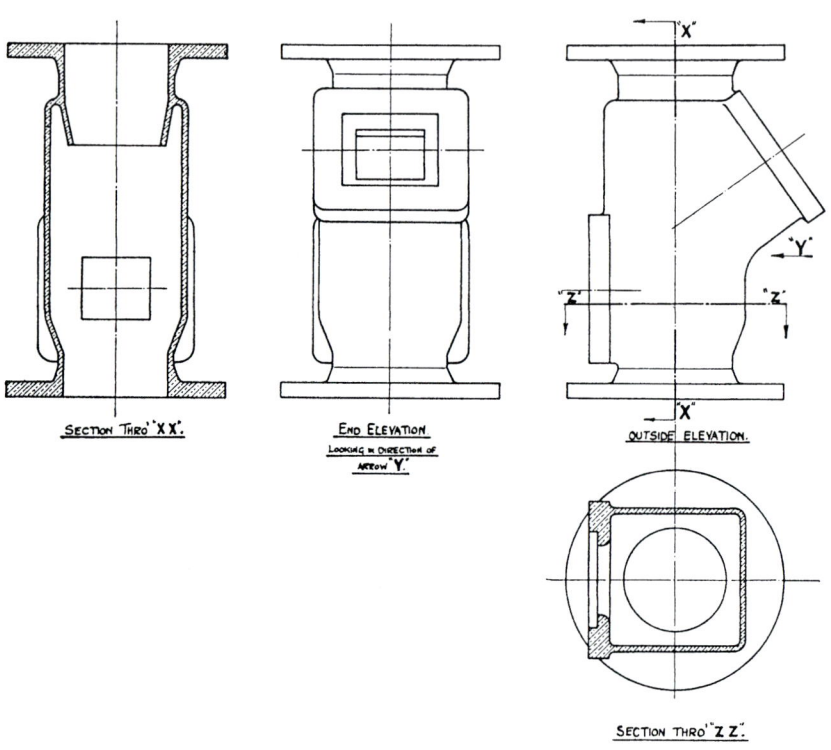

SECTION THRO' "X X".

END ELEVATION
LOOKING IN DIRECTION OF
ARROW "Y".

OUTSIDE ELEVATION.

SECTION THRO' "Z Z".

(7)

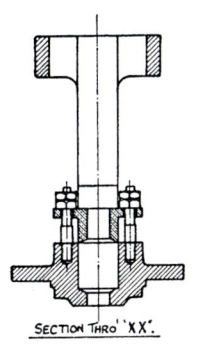

SECTION THRO' 'XX'.

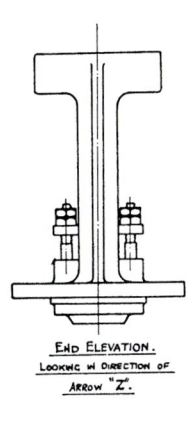

END ELEVATION.
LOOKING IN DIRECTION OF
ARROW "Z".

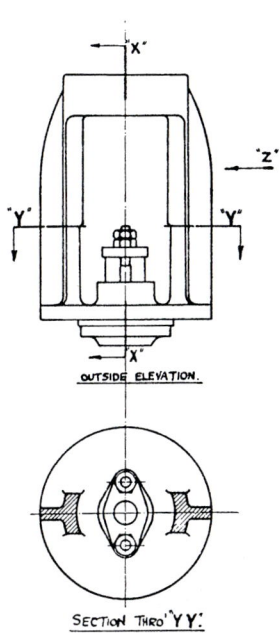

OUTSIDE ELEVATION.

SECTION THRO' YY.

(8)

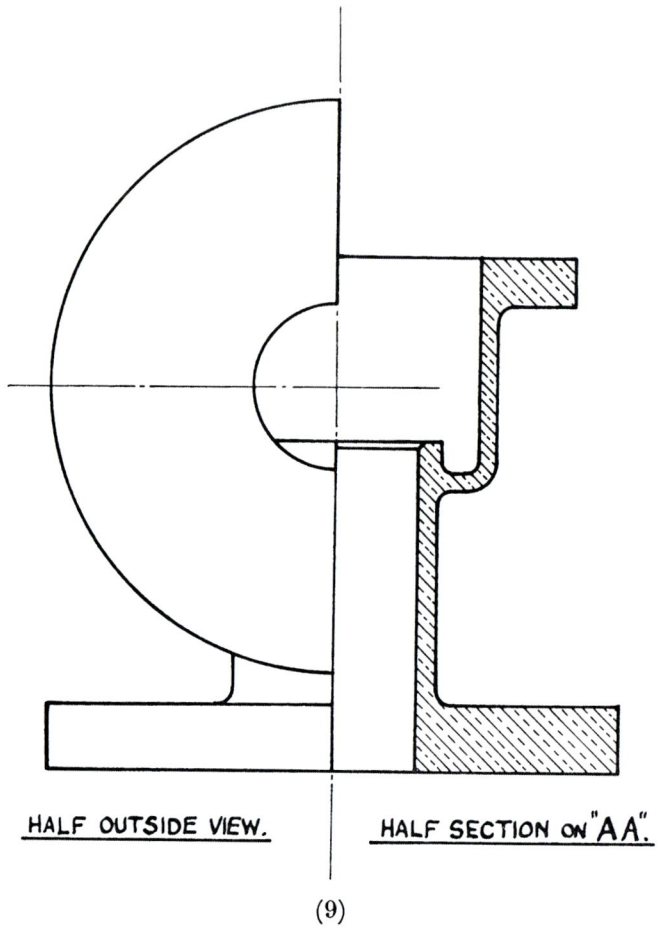

HALF OUTSIDE VIEW.   HALF SECTION ON "AA".

(9)